The Quest for Physical Theory

George A. Reisch is the author of *How the Cold War Transformed Philosophy of Science* (Cambridge, 2005), *The Politics of Paradigms: Thomas S. Kuhn, James B. Conant, and the Cold War "Struggle for Men's Minds"* (SUNY Press, 2019), and coeditor of *The Humanistic Background of Science*, by Philipp Frank (SUNY Press, 2021).

The Quest for Physical Theory
Problems in the Methodology of Scientific Research

by Thomas S. Kuhn

Thomas S. Kuhn's Lowell Lectures

Delivered in March 1951

Edited with an Introduction by
George A. Reisch

Library of Congress Control Number: 2021911436

© 2021 George A. Reisch. Rights to Thomas Kuhn's Lowell Lectures are adminstrated by MIT Libraries Department of Distinctive Collections. Except for scholarly quotation, no part of this edition may be reproduced without permission.

FREE PUBLIC LECTURES IN THE
LOWELL INSTITUTE
(Founded by John Lowell, Jr., in 1836, and opened to the public in 1839)

THE QUEST FOR PHYSICAL THEORY
Problems in the Methodology of Scientific Research

By THOMAS S. KUHN, A.M., Ph.D.
Junior Fellow, Society of Fellows, Harvard University

1. FRIDAY, MARCH 2
INTRODUCTION: TEXTBOOK SCIENCE AND CREATIVE SCIENCE

2. TUESDAY, MARCH 6
THE FOUNDATION OF DYNAMICS

3. FRIDAY, MARCH 9
THE PREVALENCE OF ATOMS

4. TUESDAY, MARCH 13
"THE PRINCIPLE OF PLENTITUDE": SUBTLE FLUIDS AND PHYSICAL FIELDS

5. FRIDAY, MARCH 16
EVIDENCE AND EXPLANATION

6. TUESDAY, MARCH 20
COHERENCE AND SCIENTIFIC VISION

7. TUESDAY, MARCH 27
THE ROLE OF FORMALISM

8. FRIDAY, MARCH 30
CANONS OF CONSTRUCTIVE RESEARCH

To be given in the Lecture Hall of the Boston Public Library, Copley Square, at 8 P.M. Doors opened at 7:30 P.M., but closed at 8 o'clock and THROUGHOUT EACH LECTURE.
FREE TICKETS may be obtained by writing to THE CURATOR OF THE LOWELL INSTITUTE, BOSTON PUBLIC LIBRARY, COPLEY SQUARE, BOSTON, 17.

Flyer circulated by the Lowell Institute to advertise *The Quest for Physical Theory*, side 2.

Thomas Kuhn's *Quest for Physical Theory*: Editor's Introduction

By George A. Reisch

After its publication in 1962, *The Structure of Scientific Revolutions* quickly established Thomas Kuhn as one of America's most influential historians and philosophers of science.[1] But in 1950, when he received an invitation from Boston's Lowell Institute to join the ranks of its distinguished lecturers, he had just started teaching at Harvard University and was largely unknown outside his circle of colleagues and friends. The incongruity was not lost on Kuhn. Upon greeting his audience at the Boston Public Library, he humbly acknowledged the "eminent scholars who have preceded me on this platform" and his hope to repay them "a small portion of my indebtedness."

Since it was founded in the 1830s by John Lowell Jr., the institute had tasked leading scholars from around the world to promote "the moral, intellectual, and physical instruction and education of the inhabitants of Boston." The historians, artists, linguists, statesmen, scientists, and philosophers who accepted this commission enjoyed generous stipends, the opportunity to hold forth at great length—Kuhn was offered the choice of six or eight one-hour lectures—and the sustained attention of eager, self-selecting audiences. Tickets were free to anyone. But they had to be requested in advance and—it was well known—one had to arrive on time. Doors closed promptly at eight o'clock and would not reopen for any whose tardiness would postpone or disrupt the edification of those inside.[2]

In science, there had been the geologist Charles Lyell, the biologist Julian Huxley, the naturalist Louis Agassiz, and the astronomer Harlow Shapley. The British philosophers Alfred North Whitehead and Bertrand Russell and the Americans Josiah Royce, Charles Sanders Peirce, and William James had preceded Kuhn, as well. Whitehead's book *Science and the Modern World* and Russell's *Our Knowledge of the External World* were first written as Lowell lectures, as was James's *Lectures on Pragmatism*—a series said to have concluded in 1907 with a standing ovation.[3] The lectures also helped to launch the careers of scholars who belonged to Harvard's Society of Fellows, itself founded and endowed by Abbott Lawrence Lowell in 1933 upon his retirement as President of Harvard. The historian Arthur Schlesinger Jr., for example, used his lectures to preview his forthcoming book, *The Age of Jackson*, for which he would later win a Pulitzer Prize at the age of twenty-nine.

When Kuhn received his invitation, he was twenty-eight years old and, like Schlesinger, a Junior Fellow in the Society. Despite his humble opening remarks, however, Kuhn was ambitious and determined to distinguish himself.[4] In the weeks and months before his lectures, as he outlined and sketched what he planned to say, he envisioned nothing less than a revolution in our understanding of science. His finished lectures brim with confident, far-reaching claims about the fundamental levers and springs of human inquiry, the nature of scientific knowledge, and why other theories of science miss their mark. Kuhn also planned to follow Whitehead, James, and Russell by publishing his lectures as a book—one that would transform our understanding of science and inaugurate a new, interdisciplinary field of research.[5]

Yet the lectures Kuhn delivered on Tuesdays and Fridays in March, 1951 were not *The Structure of Scientific Revolutions*. They were not even published. To scholars thumbing through the personal files that Kuhn left to MIT after his death in 1996, it may have seemed natural that the famous author of *Structure* had once been so honored by the Lowell Institute. But the lectures themselves remind us that Kuhn's now-classic theory of science took shape slowly, beginning in the late 1940s and continuing through the 1950s. The Lowell Institute provided Kuhn an early opportunity to systematize and present his emerging theory, but the resulting lectures show his ideas and terminology still in flux. The youthful, revolutionary confidence Kuhn invested in the lectures was not misplaced; but it bore fruit only later—after he had refined and strengthened his ideas, the terms in which he expressed them (notably, the "paradigms" of *Structure* do not appear in the Lowell lectures), and after his skills as a teacher and writer had matured and stabilized. From this wider perspective, the lectures speak to the development of Kuhn's ideas,

Contents

Thomas Kuhn's
Quest for Physical Theory: Editor's Introduction   xi

I
Introduction: Textbook
Science and Creative Science   1

II
The Foundations of Dynamics   21

III
The Prevalence of Atoms   41

IV
'The Principle of Plenitude':
Subtle Fluids and Physical Fields   57

V
Evidence and Explanation   77

VI
Coherence and Scientific Vision   97

VII
The Role of Formalism   117

VIII
Canons of Constructive Research   137

Appendix
The Prevalence of Atoms:
Kuhn's Original Outline   153

What are the problems of  scientific research today

In a world in which science's quest for physical theory has already had results that promise to change the course of history, the fate of mankind may depend upon solving the problems of research . . .

The Lowell Institute

presents a course of eight free lectures on "The Quest for Physical Theory" by Thomas S. Kuhn, A.M., Ph.D. — Tuesdays and Fridays at 8 p.m., beginning Friday, March 2 (omitting Friday, March 23), at the Boston Public Library.

[over]

Flyer circulated by the Lowell Institute to advertise *The Quest for Physical Theory*, side 1.

the influences behind them, and the postwar intellectual climate in which they developed.

Thomas Kuhn's Path to the Lowell Lectures

Born in Cincinnati, Thomas Kuhn was raised in New York City and attended preparatory schools in Pennsylvania and Connecticut. He enrolled at Harvard in the fall of 1940, graduated in 1943 with a degree in physics, and served in the second world war as a radar specialist. He subsequently returned to Harvard for graduate work in physics and earned his Ph.D. in 1948.

Several factors may have helped to distinguish Kuhn in the eyes of Ralph Lowell, the trustee who invited him to lecture in the 1950-51 season.[6] Besides Kuhn's membership in the Society of Fellows, Lowell enjoyed a working relationship with James Bryant Conant, the President of Harvard who had succeeded Abbott Lawrence Lowell—Ralph's cousin—in 1933. As President, Conant managed the university's longstanding relationship with the Lowell family, whose generations of politicians, businessmen, scientists, and intellectuals contributed to the industrial growth and cultural prestige of the Boston area. It was Conant, for example, who suggested to Ralph Lowell that the institute produce radio programs featuring scholarly lectures and round-table discussions with faculty from local colleges and universities. By 1951, the year Kuhn delivered his lectures, the institute had created WGBH, which grew to become the public radio and television station it remains today.[7]

Ralph Lowell shared Conant's view that postwar economic growth and rapid cultural change required Americans to better understand their world. That included modern science which was sure to shape the domestic and geopolitical challenges that lay ahead. Few had given these challenges more thought than Conant, who had helped lead the Manhattan Project's development of the atomic bomb. Lecturing at Yale in 1946, Conant put it this way:

> We need a widespread understanding of science in this country, for only thus can science be assimilated into our secular cultural pattern. When that has been achieved, we shall be one step nearer the goal which we now desire so earnestly, a unified, coherent culture suitable for our American democracy in this new age of machines and experts.

A strong, vibrant democracy, he believed, would require most educated Americans—certainly civic and institutional leaders—to have a basic understanding of how modern science works. For the vast majority whose educations would not include advanced classes or research experience, Conant recommended learning the ways of science through history—through detailed case studies of momentous experiments and debates of the past, such as those surrounding the discovery of oxygen, of air pressure, or Copernican astronomy. Without setting foot in a research lab, Conant believed, they would become citizens better equipped to understand the complex, sometimes accidental, ways by which knowledge grows and "science moves ahead."[8]

To spearhead this national effort, Conant initiated a new program in general education at Harvard that Thomas Kuhn joined as he concluded his graduate work in physics. Alongside Conant, Kuhn taught historical case studies to future bankers, lawyers, writers, and senators. He also represented Harvard's program at national conferences dedicated to general education and techniques for teaching science to ordinary citizens.

Whether Conant had personally recommended Kuhn to Ralph Lowell, or whether it was his membership in the Society of Fellows that brought him to Lowell's attention, Conant prepared the ground for Kuhn's invitation. For Conant had nominated Kuhn for membership in the Society shortly after they embarked on what would become a long, fruitful intellectual friendship and (at times) rivalry. In Kuhn, Conant found a protégé whose talents and interests complemented his plans to educate the public about science—not only through courses in general education for undergrads at Harvard and other colleges, but through semipopular books such as Conant's *On Understanding Science* and *Science and Common Sense*. In his foreword to Kuhn's first book, *The Copernican Revolution* of 1957, Conant applauded its effort to enlighten the educated public and chart "the road which must be followed if science is to be assimilated into the culture of our times."[9] In Conant, Kuhn found a mentor who offered him a much-desired career path outside of physics, whose intellectual passion for the history and philosophy of science rivaled his own, and who seeded many of the ideas that circulate through Kuhn's Lowell lectures and later grew into *The Structure of Scientific Revolutions*. As Kuhn declared in the dedication to its first edition, it was "James Bryant Conant/who started it."

The Quest for Physical Theory: Foundations, Goals, and Organization

Kuhn gladly accepted Ralph Lowell's invitation and proposed to deliver eight lectures about "several aspects of scientific behavior, or method, as they may be abstracted from the history of scientific ideas."[10] Inspired by what he later described as his "Aristotle experience"—a specific moment, likely in the summer of 1947, when he found to his surprise that Aristotle's long-discarded theories of nature seemed sensible and convincing when apprehended in the right ways—Kuhn had been using his time as a Junior Fellow to read widely in several fields. The "several aspects" of science he planned to cover included history, philosophy, psychology, logic, semantics, and linguistics, all of which Kuhn enlisted to help explain how theories like Aristotle's, coherent and successful on their own terms, were superseded and remembered simply as mistaken or false. However this process worked, it seemed clear to Kuhn that science did not change as textbooks and popularizations typically claimed—by replacing false with true theories and increasing humanity's stock of knowledge. Instead, science moves forward by embracing different, but not necessarily better, theories. And as it does, it leaves some knowledge behind—like the theories of Aristotle that startled Kuhn with their intrinsic logic and sensibility.[11]

Kuhn was particularly interested in Freudian psychology and its view that powerful ideas and drives remain invisible and unknown to the conscious mind. If unconscious complexes could be brought to light through psychoanalysis (which Kuhn himself underwent in the late 1940s), then perhaps methodological analysis could reveal unconscious or forgotten factors in science that shaped its history and its understanding of its past.[12] As a child and through his undergraduate years at Harvard, Kuhn was also deeply interested in politics, political theory, and—in the wake of Germany's stunning collapse into totalitarianism in the 1930s—the seeming powers of propaganda and ideology to influence the human mind.[13]

Not long before Lowell's invitation arrived, it appears, Kuhn found a kindred intellectual spirit in the Polish physician Ludwik Fleck. Fleck's book *Entstehung und Entwicklung einer wissenschaftliche Tatsache* of 1935, later translated and published as *Genesis and Development of a Scientific Fact*, confirmed and encouraged Kuhn's theorizing about the social, psychological, and ideological dimensions of scientific thought and method.[14] Fleck agreed that science functioned as it did not because its beliefs reflected objective truths of nature, but because they supported a sociological *Denkstil*, or style of thought, and a collective "harmony of illusions" among scientists.

One illustration of Fleck's influence on Kuhn is the title he first chose for his lecture series. He wrote to Ralph Lowell, "The series might appropriately be titled 'The Creation of Scientific Objects'"—a title that echoes Fleck's and likewise subverts ideals of objectivity and realism by promising to uncover the creative origins of knowledge. Kuhn soon chose a different title, however: "The Quest for Physical Theory: Problems in the Methodology of Scientific Research."[15] If this title suggests a more traditional image of research as discovery—as opposed to creation—of knowledge, Kuhn made it clear in his first lecture that this was not his intention. The objects studied by modern science, the facts we learn about them, and the laws that govern them are not passively discovered but creatively forged in the course of fruitful research.

A foundation on which Kuhn rested this insight was a notion he had embraced at least since his undergraduate days: the "infinite" complexity of experience from which our knowledge is forged.[16] In his lectures, he invokes William James's memorable description of infant experience as a "bloomin, buzzin, confusion" to argue that science becomes possible only when this original chaos of experience is reduced and simplified; when those parts that seem important, interesting, or puzzling from within some framework or conceptual scheme are selected and attended to—while the rest of experience is ignored. Readers familiar with *Structure* may identify these frameworks with Kuhn's paradigms. But in 1951 Kuhn had not yet formulated his theory of paradigms. The word appears a few times, but only in its traditional pedagogic sense—not to explain the foundations of science and the structure of its historical development.[17]

Instead of paradigms, *The Quest for Physical Theory* appeals to an array of theoretical devices or constructs that serve to reduce and focus experience. Over the course of his eight lectures, Kuhn calls them "preconceptions," "prejudices," "theories," "points of view," "orientations," "metaphors," "meaning systems," "mental frameworks," and "behavioral worlds." They drive science forward by shaping how scientists think and reason, how they perceive nature and events, and by organizing and reducing the welter of experience to help ensure that scientists are not distracted from their research by novelties.

If all of this seemed unfamiliar to Kuhn's audience, he reassured them that the ways of science are indeed quite different from what they may have learned in school or from popular books. As Conant had in his book *On Understanding Science*, he singled out Karl Pearson's *The Grammar of Science*, of 1892, for its portrayal of scientists as unbiased, self-effacing recorders of natural phenomena. This is misleading, Kuhn explained, because it is impossible:

> I should like to suggest that the impartial, dispassionate observation of nature is impossible, that there are no "pure facts" from which alone valid theories can be derived, and that the effort toward "self elimination" which Pearson proposes as the scientist's goal, would, in practice, result in the abolition of productive research. (Lecture I)

In fact, Kuhn explained later, research is deeply subjective and personal. So much so, the methodology of science is circular and self-reinforcing as empirical evidence and the theoretical conclusions it supports remain merged within these all-important "orientations" or "points of view":

> This suggests that scientific research is inherently circular, that it does not proceed from experimental facts to theories, but that facts and theories are provided together, in a more or less inchoate form, by scientific orientations. (Lecture V)

Research that unfolds within the confines of a particular "orientation," Kuhn explained, is "textbook science," which constitutes a settled body of knowledge. This is very different from "creative science" in which these orientations themselves are realized and refined. Even talented scientists, Kuhn believed, are ill equipped to investigate and understand creative science because it is driven by complex, historical processes typically overlooked by textbooks. Only the methodologist, equipped with the tools of logic, philosophy, history, and psychology can discern and disentangle the factors in play.

Kuhn dedicated his first lecture to this basic distinction between textbook and creative science. The distinction anticipates *Structure*'s opening remarks about "the textbooks from which each new scientific generation learns to practice its trade" and the "decisive transformation" in store for those willing to consider history and the creative, sociological dynamics of research.[18] In *Structure*, this distinction would become fundamental to important differences between "normal" and "revolutionary" research and the historical cycle through which "normal science" is beset by "anomalies," reaches a "crisis," and is finally transformed by a new, different paradigm and accordingly rewritten textbooks. In *Quest*, Kuhn outlines a similar historical process using largely different terms. It is not "normal science" but instead "the classical period of a scientific orientation" that remains stable and well represented by textbooks. Then difficulties emerge that lead to a "crisis stage" and finally a revolution through which "some alternate orientation" resolves the issues at hand. While textbooks will simplify these complexities and present them as increasing "the stock of scientific knowledge," scientists living through these transitions observe a different kind of process, for "a scientific revolution is always very nearly as destructive as it is constructive" (Lecture V).

To cover this material in eight one-hour lectures, Kuhn divided his lecture series roughly in half. The first part presents historical episodes in science that Kuhn had been teaching in the general education program and some of which would later appear in *Structure*. These include the history of dynamics ("The Foundations of Dynamics," Lecture II), atomism ("The Prevalence of Atoms," Lecture III), and theories of subtle fluids ("'The Principle of Plenitude': Subtle Fluids and Physical Fields," Lecture IV). Inspired by his revelation from Aristotle, these treatments emphasize that antique theories as well as those more recently discarded, such as phlogiston chemistry, have more integrity and value than contemporary "textbook science" would suggest.

In the last four lectures, beginning with "Evidence and Explanation" (Lecture V), Kuhn turns to philosophy, psychology, and linguistics to explore the complex, "circular" process by which scientific objects and laws emerge from the primitive "flux" of human experience. He devotes "Coherence and Scientific Vision" (Lecture VI) to the psychology of perception, and "The Role of Formalism" (Lecture VII) to logic and his criticisms of formal, logical approaches to understanding science. Here Kuhn introduces a critique of formalist philosophy of science that he concludes in the final lecture, "Canons of Constructive Research" (Lecture VIII).

The Significance of the Lectures Today

Kuhn's determination to identify and untangle the hidden sources of scientists' ideas, perspectives, and creativity naturally invites questions about the sources of his own ideas, especially those in *The Structure of Scientific Revolutions* that widely influenced postwar intellectual life. Besides his momentous encounters with Aristotle and Ludwik Fleck, the lectures amply document Kuhn's familiarity with the writings of other historians and philosophers. Lecture V, for example, shows Kuhn adapting Kantian epistemology to his methodological picture of science. Other philosophical influences at work include Kuhn's colleague Philipp Frank, an original member of the Vienna Circle of philosophers who also worked within Conant's general education program and shared Kuhn's interest in the sociology of science,[19] and the philosopher W.V.O. Quine, whom Kuhn knew in the Society of Fellows. Quine's critique of philosophical formalism in his classic essays "On What There Is" and "Truth By Convention" can be heard clearly in Kuhn's discussions of "meaning systems" and the roles they play in research. Quine's soon-to-be-famous essay "Two Dogmas of Empiricism," which Kuhn may have read before its publication in 1951, has at least two points of contact with Kuhn's lectures: Quine's own emphasis on the "flux" and "barrage of sensory experience" and Kuhn's remark in Lecture V that "science is an interlocking fabric."[20]

In history of science, Kuhn's lectures illustrate the style, and in some cases the content, of the case studies that Conant and others in the general education program were then developing and teaching, initially on the basis of Conant's book *On Understanding Science*.[21] Kuhn's view that research is driven largely by scientists' ideas illustrates his debts to the "conceptual schemes" that Conant regularly emphasized in his writings, to Annaliese Maier's studies of medieval theories of motion, and to Alexandre Koyré's studies of Galileo.[22] Kuhn's life-long interest in psychology complemented this view of science's history. In Lecture VI, he discusses the Swiss developmental psychologist Jean Piaget, whose research focused Kuhn's attention on how children sort and compare conflicting descriptions and conceptions of natural events.[23] He also discusses experiments documenting the plasticity and theory-laden qualities of perception (such as those of his Harvard colleagues Jerome Bruner and Leo Postman) that Kuhn would return to in *Structure* as evidence for the power and priority of paradigms in guiding research.

The lectures also indicate Kuhn's early interest in semantics, manifest as much in his scholarly reading of Quine and Bertrand Russell as in popular authors like Stuart Chase, author of *The Tyranny of Words*, and Benjamin Lee Whorf, whose writings on Hopi linguistic categories Kuhn borrowed in Lecture VII to argue that language influences scientists' perceptions.[24]

It is sometimes remarked that in the wake of *Structure* Kuhn's thinking took a "linguistic turn" toward semantics and away from the sociological and psychological themes within *Structure*. In light of *Quest*, however, this development appears more as a linguistic return to early theorizing about the semantics of "meaning systems" that Kuhn debuted here.[25]

Debates about Kuhn's and *Structure*'s significance for twentieth-century philosophy also stand to be informed. Kuhn's historic encounter with Karl Popper and his followers in 1965 at the colloquium on philosophy of science at Bedford College, for example, remains itself a paradigm in Kuhn scholarship.[26] The title of Kuhn's anti-Popperian lecture at that event, "The Logic of Discovery or the Psychology of Research?" suggests a relatively narrow, even tentative, interest in the psychology of science. Yet the Thomas Kuhn that spoke at the Boston Public Library more than a decade before offered a panorama of considerations that offers a broader, if not stronger, argument against Popperianism: Popper ignores not only the psychology but also the semantics, logic, education, and anthropology of modern science—simply too much for the responsible methodologist to sacrifice for the alluring logical simplicity of Popper's falsificationism.

As for logical empiricism, among the most important philosophical movements in the United States after the war, it is often remarked that *Structure*'s first sentence—"History, if viewed as more than a repository for anecdote or chronology, could produce a decisive transformation in the image of science by which we are now possessed"[27]—was directed at logical empiricism's image of theories as logical structures and of scientific research as applications of logic. On this reading, Kuhn made his case Trojan-horse style, from within the *International Encyclopedia of Unified Science*, a project conceived and established by the Viennese economist and philosopher Otto Neurath and designed to articulate and promote logical empiricism widely throughout science, the academy, and modern life. After his death in 1945, Neurath left the encyclopedia to his coeditors Rudolf Carnap, who like Neurath had belonged to the Vienna Circle, and Charles Morris, the pragmatist philosopher who recruited Kuhn to contribute to the encyclopedia in the early 1950s.

Kuhn indeed saw himself as a philosophical revolutionary. He explained to Conant on the eve of *Structure*'s publication that his account of perception (in *Structure*'s chapters 6 and 10) attacked the notion that scientific observation is "passive" (a notion that appears here in Lecture VI when Kuhn compares "the individual perceiver" to "a camera combined with a responder"). "If that notion can be licked," Kuhn explained to Conant, "the rest of the structure falls" and philosophy since Descartes faces revolutionary transformation. *Structure* also called for reform in semantics by denying the possibility of theory- or paradigm-independent observation languages (such as those taken to ground the unity of the sciences); in metaphysics by insisting that the world itself changes—at least in some vague but undeniable sense—when paradigms shift; and in historiography by subverting "positivist," accumulationist assumptions that newer theories absorb and preserve the fruitful insights of their predecessors. To those irked by *Structure*'s appeals to "faith" and "conversion experiences," Kuhn seemed to argue not only that philosophical analyses had so far failed to understand science, but that scientific revolutions are not explicable through philosophy at all.[28]

In other respects, however, the Trojan legend is misleading. For Kuhn later acknowledged that he did not understand logical empiricism well in the early 1950s, especially the mature, pragmatically inflected views of Carnap which, it was later pointed out, are not wholly different from those Kuhn presented in *Structure*. While scholars in the 1990s debated and retheorized Kuhn's relationship to logical empiricism, Kuhn himself felt a sense of tragic regret. For when Carnap approved *Structure* for inclusion in the *International Encyclopedia* and wrote to Kuhn to express his admiration for the book and the issues it raises, Kuhn chalked it up to expediency and politeness. Only in retrospect, Kuhn realized, Carnap's letter signaled that they shared substantive philosophical interests; that, as Kuhn put it, "he and I might usefully talk." Prior to Carnap's death in 1970, they could have discussed these interests, but never did.[29]

The Quest for Physical Theory adds to this story of missed connections and misunderstandings. Perhaps most striking is Kuhn's suggestion that logical empiricism—represented by the philosophical "formalism" he criticizes in Lectures VII and VIII—aimed not just to understand science with philosophical tools, but to formalize the natural, everyday language used in laboratories, research papers, textbooks, and even public, nonscientific discourse. As Kuhn put it,

> The program for the formalization of scientific language calls for the application of this technique to other portions

> of science and to the language of scientific communication. And scientific language is for this purpose taken to be the entire language in which we discuss our perceptions. In everyday terms, it is the language in which we discuss facts or events or the relations between events. It is a language used not only by the scientist but, more loosely, by the layman. And it is this language which we are now asked to formalize. (Lecture VII)

Kuhn evidently believed that logical empiricism aimed to replace natural language with formalized language. On this basis, he argued that this formalistic crusade would effectively halt scientific progress. For it is the ambiguity and complexity of natural language—manifest in what he calls "meaning fringes" around terms—that allows science to change and evolve. He explained,

> We do leave vague meaning fringes on scientific terms, and our research is always conducted within the area determined by these vaguer fringes. It is in these areas alone that questions can arise as to established theories. . . . It is only in the area provided by meaning fringes that scientific questions can arise and that scientific exploration can occur. (Lecture VIII)

The vagueness and ambiguity of language that formalists (allegedly) aimed to eradicate, in other words, is itself a necessary precondition for fruitful research and scientific progress.

In early publications (such as the Vienna Circle's philosophical manifesto *Wissenschaftliche Weltauffassung*), logical empiricists did issue bold, revolutionary claims about philosophy's power to unify the sciences and to reform and reinvigorate culture and modern life. Linguistic reforms were a part of this crusade. Carnap's essay "The Elimination of Metaphysics through the Logical Analysis of Language" and Otto Neurath's proposal to create an *index verborum prohibitorum*—a list of prohibited words—targeted specific instances of misleading, unscientific language. But neither called for the elimination of natural language and its replacement by logically precise alternatives because logical empiricists understood that natural language is an essential foundation on which scientific knowledge grows. Carnap called it the "thing language" in which higher-level observations or theories can be tested and confirmed. Neurath frequently emphasized that our "universal jargon" or "everyday language" facilitates the fruitful collaboration of scientists in different fields, even from different national cultures, in ways that substantiate the unity of the sciences. In this sense Neurath could speak of his *International Encyclopedia* becoming "a living intellectual force growing out of a living need of men, and so in turn serving humanity."[30]

Neurath also championed the vagueness, ambiguity, and complexity of scientific language. Science begins, he wrote in the encyclopedia, with what Germans and French might call a linguistic *Ballung* or *grégat*—"a full lump of irregularities and indistinctness, as our daily speech offers it." Logical empiricism's task was not to ignore or replace this indistinctness through formalization, but instead to exploit the resulting opportunities for advancing and unifying scientific knowledge. "Empiricists cannot refrain from using faint and blurred expressions with rather vague outlines,"[31] Neurath wrote in 1944—seven years before Kuhn lectured on "meaning fringes."

Whether or not Kuhn was familiar with these writings, he was familiar with two monographs in Neurath's encyclopedia, one by the American linguist Leonard Bloomfield and another by the British biologist Joseph Woodger, both of which Kuhn had read by the late 1940s.[32] Bloomfield and Woodger unmistakably exalted the power of formal, logical analysis to clarify theories and concepts and to diagnose misunderstandings and disagreements occasioned by the vagaries of language. But neither aimed to replace ordinary language.[33] How then did Kuhn come to believe that logical empiricism aimed to formalize the natural language of science and everyday life? One possibility is that he conflated logical empiricism with the general semantics movement and popular books that he mentions in his last lecture, some of which did prescribe sweeping changes in everyday habits of speech.[34]

Regardless of how Kuhn formulated his argument, it echoed a continuing tradition of philosophical criticism directed at logical empiricism and its defenders. The movement had contended with similar claims since the 1930s when some of its leading figures arrived in the United States as intellectual émigrés and some American philosophers became suspicious. This philosophical import, complete with the socialist Neurath's new *International Encyclopedia of Unified Science*, seemed to have a totalitarian bent not unlike dialectical materialism, the official philosophy of Joseph Stalin's oppressive Soviet Union, and its attendant *Great Soviet Encyclopedia* of knowledge. When John Dewey accepted Neurath's invitation to contribute to his new encyclopedia, for example, he did so from an oppositional, pluralistic stance. Any reductive "attempt to secure unity by defining the terms of all the sciences in terms of some one science," he wrote, "is doomed in advance to defeat. In the house which science might build there are many mansions." In 1939, speaking on the eve of Hitler's march into Poland, William James's student Horace Kallen was sure that Neurath and his fellow editors at the new encyclopedia of science would cultivate theoretical unity not in any democratic way; they would "impose it by *force majeure*."[35]

After the war, as American scientists and intellectuals debated government's proper roles in funding and organizing postwar research, Neurath's cousin and champion, the *New York Times* science writer Waldemar Kaempffert, argued that experts in science should plan and guide national research projects for the public good. Earlier, Kaempffert had praised Neurath's new encyclopedia as a forum for scientific coordination of this sort; and he now argued that the Manhattan Project itself illustrated what modern science can accomplish quickly and effectively when it is properly funded and organized. Absolutely not, Conant and other scientific leaders replied as they attacked Kaempffert and lobbied Washington to adopt the hands-off approach to research that ultimately won the day. Top-down guidance of any kind, they argued, from would-be dictators or philosophical theories claiming to facilitate or advance research, would condemn science to failure.[36]

That *The Quest for Physical Theory* culminates in lectures VII and VIII with Kuhn's own warnings that logical empiricism would harm research suggests the extent to which Conant and this postwar debate about scientific method weighed on his early theorizing. It also points to what is perhaps the broadest significance Kuhn's lectures have today for the history of philosophy. Alongside their relevance for understanding *Structure*, in particular, they offer new documentation of this formative midcentury encounter between American and European philosophy of science. Along with memoires and essays from young Americans (such as Morris and Quine) who travelled to Europe to learn and report about exciting new trends, and along with institutions such as Neurath's encyclopedia at the University of Chicago Press and Philipp Frank's Institute for the Unity of Science in Boston, Kuhn's lectures illustrate the dynamic, exciting contrasts and affinities between American pragmatism, logical empiricism, and other programs that swirled around the future author of *The Structure of Scientific Revolutions*. Conant himself confessed to something like philosophical intoxication in this climate when he described his sensibilities as a "mixture of William James's *Pragmatism* and the logical empiricism of the Vienna circle, with at least two jiggers of pure skepticism."[37] We ought not be surprised, therefore, to find confusions as well as revealing points of contact in Kuhn's early theorizing with these two philosophical traditions.

On the pragmatic side of this encounter, *The Quest for Physical Theory* shows the young Thomas Kuhn taking for granted Dewey's view that knowledge is the fruit of social and individual problem solving, rooted ultimately in the exigencies of social and political life (Dewey's *Reconstruction in Philosophy* is listed as "read in toto" in Kuhn's 1949 reading list). Kuhn would later distance himself from Dewey's socially engaged view of inquiry, however. Though problem solving is central to *Structure*'s account of normal science, the problems Kuhn had in mind are esoteric, internal to professional scientific communities, and isolated from political, social, and cultural pressures. The influence of William James, on the other hand, arguably grew as Kuhn's theorizing developed from *Quest* to *Structure*. In both projects, Kuhn invokes James's insights about the "bloomin, buzzin confusion" of original, unorganized experience. But in *Structure* Kuhn would elaborate a theory of scientific revolutions and paradigm shifts as personal conversion experiences that points as much to Kuhn's epiphany when reading Aristotle as to James's *The Varieties of Religious Experience*—a study in the personally transformative power of ideas, faith, and commitment that Kuhn read in 1943 and praised in his notes as "A fine & truly beautiful book."[38]

Why Didn't Kuhn Publish *The Quest for Physical Theory*?

The Quest for Physical Theory remained unpublished for a number of likely reasons. One is that Kuhn was unhappy with the lectures and later characterized them as "not very good."[39] It is not difficult to see their shortcomings. Besides some recklessly provocative and exaggerated claims, such as his remark that research is "inherently circular," and dubious readings of historical episodes in science,[40] the lectures seem poorly matched to their general audience. Those who attended were most likely not professional historians or philosophers but educated citizens of Boston. In the 1930s, at least, these audiences included a reliable coterie of "loyal listeners who attend out of habit, and most of whom were elderly spinsters."[41] If the lecturers who preceded Kuhn in the 1950-

51 season are an indication—Charles F.O. Clarke of the British Broadcasting Corporation, Langdon Warner, the Curator of Oriental Art at the Fogg Museum, the CIA official Cord Meyer Jr., and the Harvard botanist Karl Sax—Kuhn's audience probably assumed that *The Quest for Physical Theory* would address contemporary developments in physics, much as these lecturers had addressed contemporary culture and postwar geopolitics. The Lowell Institute's advance publicity likewise suggested that Kuhn would discuss "the problems of research today."[42]

Kuhn made it clear in his first lecture that this was not his goal. Among other reasons, he explained, the field of methodology is not well-suited to exploring contemporary science and technology. Still, this disclaimer is unlikely to have prepared loyal listeners of any age for Kuhn's detailed historical surveys and his brisk excursions into epistemology, metaphysics, logic, semantics, and psychology.

At times, to be sure, Kuhn's presentation sparkles, such as when he describes an ordinary (albeit unnerving) experience to motivate a discussion of epistemology and perceptual psychology:

> All of you have at one time or another awakened from a dream to discover a strange, threatening figure crouching in the corner of the room. Yet another look convinced you that the threatening figure was really the familiar easy chair with a quilt piled on it. Then the figure lost its threat and you laughed at having been fooled. But were you really fooled? (Lecture VI)

Just as often, however, Kuhn turns to scholarly texts, obscure historical figures, or physical phenomena with little or no introduction.[43] Given his experience teaching science to nonscientists, it is perhaps surprising that Kuhn did not more effectively match his lectures to his audience's expectations and background knowledge. He rarely and briefly circles back from the esoteric discussions in his later lectures to his comparably lucid and engaging introduction and historical surveys; and he concludes his final lecture without offering a helpful summation or condensed thesis suitable for a public audience. While readers today can readily glimpse the outlines of *Structure* in his final paragraphs, Kuhn left his audience with abstract and tentative remarks about language, experience, perceptions, and scientific research. That these relationships seemed shifting and fluid, he noted, "is the source of the difficulty which we have encountered again and again during the course of these lectures."

Looking back at the lectures in the 1980s or '90s, Kuhn may also have seen them as culturally anachronistic and ill-suited for publication. While judicious editing might have readily transformed scientists and researchers from "men" into "women and men," the lectures are in other ways bound to the early 1950s. In this world, racial stereotypes such as, "to a Caucasian, most Chinese look alike"—an observation that Kuhn discusses in some detail in Lecture VI—were as uncontroversial as the assumption that professional scientists are uniformly male.

In a personal sense, Kuhn wrote and delivered the lectures during difficult times. He described himself in retrospect as "clearly a neurotic, insecure young man" who suffered from tremendous anxieties. Would he succeed in pioneering this new field of methodological research he described? Would he establish himself as a tenured professor at Harvard, as he hoped? "I was one of those people who was at least in real danger of breaking up because Harvard didn't want them there," he recalled apropos of his eventual failure to gain tenure and his move to the University of California at Berkeley in 1956. Kuhn wrote *The Quest for Physical Theory* during these years and later recalled, "I had a dreadful time preparing it, and I nearly cracked up."[44]

The personal anxieties Kuhn associated with the lectures may well have been connected to the politics of the McCarthy era, as well. At a time when "a subtle, creeping paralysis of freedom of thought and speech is attacking college campuses in many parts of the country," the New York Times reported, scholars increasingly avoided controversial political issues, controversial words and concepts (including "liberal," "peace," and "freedom"), and were prone to "an unusual amount of seriocomic joking about this or that official investigating committee 'getting you'."[45]

Kuhn seemed familiar with these worries. Weeks before his lectures commenced, for example, he was deeply upset by the Lowell Institute's publicity suggesting that his lectures would address contemporary science. Their flyers and advertisements not only failed to convey the historical dimensions of Kuhn's project—they positioned Kuhn as an authority on planned, socially beneficial research (it is modern science, one flyer implied, on which "the fate of mankind may depend"). Advocates of planned research were a generally

Marxist tribe that Senator Joseph McCarthy and other anticommunists were indeed out to get.

Kuhn disavowed this advanced publicity in his introductory lecture. But political anxieties resurface in his fifth lecture when Kuhn briefly alludes to the Marxist notion that changes in science may sometimes "proceed from changes in economic structure which alter scientific motivation." He then jokes, "Attention Senator McCarthy." At some point, however, Kuhn drew a line through these three words in his typescript. He did not shy away from contemporary politics entirely, for McCarthyism comes up a third time in his final lecture on semantics (to illustrate the potential for "grave injury" of loose, misused words). But he evidently decided against making this public joke about the then-infamous senator from Wisconsin.[46]

In terms of Kuhn's scholarly career, however, the most likely reason he did not publish *The Quest for Physical Theory* lay not in the 1950s but the 1960s with the enormous, tumultuous reception of *Structure*. As he noted in his introductory lecture, *The Quest for Physical Theory* was a work in progress, a reflection of "a continuing research program," and not "a report on the outcome of a completed study." This tentative posture affected the quality and consistency of the lectures, especially when Kuhn reorganizes his presentation midstream, seems unsure of what to conclude from certain considerations, or revises his terminology. This is perhaps what Kuhn meant when he later characterized the lectures as "not very good." But it also places them in the shadow of *Structure* as an immature text. As he later remarked, *The Quest for Physical Theory* can be understood as a preliminary attempt to write *The Structure of Scientific Revolutions*.[47]

The two works are similar enough for Kuhn to have reasonably expected that *Quest*'s publication would reignite the contentious debates and criticisms that engulfed *Structure*—about what "paradigm" really means, about the very rationality of science, and the reality (or not) of scientific progress. Kuhn found these debates confused and frustrating. As suggested by his repudiation of the term "paradigm" at the end of the 1960s, and his scant use of *Structure*'s terminology in his subsequent book on the history of quantum theory, Kuhn wished to avoid these debates.[48]

For contemporary readers and scholars, however, *Quest* and its differences from *Structure* shed valuable light on Kuhn's theorizing and his intellectual debts to Conant and others. Compared to *Structure*, for instance, the lectures adopt a less sociological, more individualistic view of scientists, of the "orientations" or "behavioral worlds" in which they live, and how these psychological dispensations function to guide and shape research. The "classical period" in which these individuals conduct their research is also less restrictive than *Structure*'s "normal science," in which shared educations, paradigms, and sociological codes conform individuals to their professional community and its dogmas.[49] This dogmatism lays necessary groundwork for *Structure*'s theory of revolutions as paradigm shifts and for its related claims about incommensurability and progress. But dogma and revolution are less conspicuous in *Quest* with its more continuous—and Conantian—picture of science that "moves ahead" from one conceptual scheme to another.

To adapt Kuhn's introductory distinction between creative and textbook science, *The Quest for Physical Theory* opens a window onto the creative background to *Structure*, before it became a classic text for the historical study as well as the popular understanding of science. Where *Structure* initially promised "a decisive transformation" in our understanding of science and its history, *The Quest for Physical Theory* may at least broaden, if not transform, our understanding of Kuhn and his celebrated theories of science. It illuminates the contexts, goals, anxieties, and historical accidents that shaped his ideas and may suggest new questions about how and why Kuhn's ideas took hold of the scholarly and popular imagination—successfully and enduringly in the pages of *Structure*, but less so in the Boston Public Library in the winter of 1951.

⤴

The lectures are held in Kuhn's archived papers at the MIT Libraries Department of Distinctive Collections (collection MC-0240). Their multiple paginations in Kuhn's hand suggests that after typing a first draft of each lecture, he edited and retyped individual sections, sometimes repeatedly. He additionally annotated these typed pages by hand, sometimes deleting or adding words, sentences, or paragraphs. He bracketed portions of text, some of which he marked as optional or to be omitted if his time were running short. In cases where the surviving pages include multiple rewrites, I have preserved those revisions that appear to be final, as well as all bracketed sections. Exceptions, such as when preliminary choices of words or phrasings seem to be of potential interest, are pointed out in footnotes. Kuhn's original pages include short parenthetical references, such as "(Pearson 9)" that appear in his main text. I have removed these from the main text and placed them in quotation marks at the head of the editorial notes which appear at the end of each lecture.

Though Kuhn's typed pages contain no graphical elements, it is clear from his text that he frequently directed himself to a blackboard to elaborate or illustrate points for his audience. On several of these occasions I have inserted tables, figures, or diagrams to illustrate the relevant parts of Kuhn's discussion.

I have found no additional recordings, reports, or reviews of the lectures that indicate whether Kuhn read from these pages or improvised on their basis. With one exception, discussed below, he prepared them and referred to them as "scripts" that could have been read to his audiences word for word, complete with asides and even occasional jokes.[50] If and until additional information emerges about what Kuhn did or did not say during each lecture, the lectures presented here can at least be trusted as authentic documents of his research, his theorizing, and how he envisioned presenting them to his audience.

The exception is the third lecture, "The Prevalence of Atoms," which exists only in the form of a typed outline. Although there exists a rough outline for the entire series of lectures, none of the other lectures exists in outline form in Kuhn's files. This suggests that he adopted a different approach to preparing and delivering this particular lecture. Instead of a readable script, it appears, this outline would guide a more extemporaneous, casual, and improvisational lecture. This interpretation is supported by hand-written notes on the outline, similar to those on his fully prepared scripts, such as "Omit if past 20 min." or "read" next to important sections to be read verbatim to his audience. It is also supported by Kuhn's memories of his "dreadful time" preparing the lectures. Speaking more generally about his teaching in the late 1940s and early '50s, he contrasted two methods: he would "just go in with rough notes—knowing [that] I knew the stuff—and start talking," or he would prepare lectures thoroughly. Ironically, he recalled, when taking this careful approach Kuhn found himself "spending too much time preparing, getting very nervous in advance."[51] The unique form of the third lecture suggests that Kuhn vacillated between these two methods for his Lowell lectures, as well. After preparing scripts for the first two, he resolved to be more improvisational for the third but reverted to his original method for the remaining five.

Fortunately, Kuhn's outline for "The Prevalence of Atoms" contains an abundance of detailed information that I have elaborated and expanded into a readable script. This version preserves the many complete sentences and phrases in Kuhn's outline and expands more telegraphic points and asides into complete sentences and paragraphs. For readers interested in the outline itself, it is reproduced here as an appendix.[52]

Notes

1. Thomas Kuhn, *The Structure of Scientific Revolutions*, 4th ed. (Chicago: University of Chicago Press, 2012), hereafter *Structure*.

2. According to the institute's original charter, quoted in the *Boston Globe*. The article added, "You may be late for a play, the opera, a ballet—even church. But once a Lowell Institute lecture begins, the doors are closed. Just try to get in—it's impossible." "You Have to Be on Time for a Lowell Institute Lecture," *Boston Globe*, Jan 28, 1951.

3. On the reception of James's lectures, see Linda Simon, *Genuine Reality: A Life of William James* (New York: Harcourt, Brace, and Company, 1998), 351. In his "Outline for Lowell Lectures, Spring 1951, TSK 3-30-1950," Kuhn refers to Whitehead's and James's books (Thomas S. Kuhn Papers, Department of Distinctive Collections, MIT Libraries [hereafter: TSK-MIT], box 3, folder 10).

4. On Kuhn's intellectual ambitions, see for example Karl Hufbauer, "From Student of Physics to Historian of Science: T.S. Kuhn's Education and Early Career," *Physical Perspectives* 14 (2012): 421–70.

5. Kuhn to Professor David Owen, Jan. 6, 1951, TSK-MIT, box 3, folder 10. In this letter Kuhn describes this new field of research, his plan to elaborate it in his then upcoming Lowell lectures, and the likelihood that he will publish this research as a book.

6. Kuhn's invitation from Ralph Lowell is mentioned in Kuhn's correspondence with William H. Lawrence (William H. Lawrence to Thomas Kuhn, April 1, 1950, Lowell Institute Records, Massachusetts Historical Society, carton 10).

7. Edward Weeks, *The Lowells and Their Institute* (Boston: Little, Brown, 1966), 143, 163–71.

8. James Bryant Conant, *On Understanding Science: A Historical Approach* (New Haven: Yale University Press, 1947), 3, 25. For an overview of Conant's vision of public history of science and professional academic resistance to it, see Christopher Hamlin, "The Pedagogical Roots of the History of Science: Revisiting the Vision of James Bryant Conant," *Isis* 107(2) (2016): 282–308.

9. James Bryant Conant, "Foreword," in Thomas S. Kuhn, *The Copernican Revolution* (Cambridge, MA: Harvard University Press, 1957), xviii. James Bryant Conant,

Science and Common Sense (New Haven: Yale University Press, 1951). In correspondence from 1950, Conant speaks to Kuhn of their "joint efforts" to develop case studies and describes his book *Science and Common Sense* as "my latest excursion into popularizing the methods of science." On one occasion, at least, Kuhn offered to Conant results of his current research on Galileo and Torricelli that he deemed appropriate to be included "in the new edition of OUS [*On Understanding Science*] or in anything else which you have underway" (Kuhn to Conant, Oct. 8, 1950; Conant to Kuhn, Oct. 11, 1950, Harvard University Archives, Records of James B. Conant, Box 402). Courtesy of Harvard University Archives. Kuhn noted that Society fellows were often invited by the Lowell Institute in *The Road Since Structure* (Chicago: University of Chicago Press, 2000), 289.

10. Kuhn to Ralph Lowell, March 19, 1950, TSK-MIT, box 3, folder 10.

11. For more on Kuhn's "Aristotle Experience" and its place in his career, see my essay "Aristotle in the Cold War" in *Kuhn's* Structure *at Fifty*, ed. R.J. Richards and L. Daston (Chicago: University of Chicago Press, 2016), 12–29.

12. In his early outlines for the lectures, Kuhn noted "Fleck or Freud, etc." as possible illustrations of "enlightenment due to a changed viewpoint." "Outline for Lowell Lectures, Spring 1951, TSK 3-30-1950," TSK-MIT, box 3, folder 10. The annotation appears on p. 7 of these notes. For more on Kuhn's interest in psychoanalysis, see my "On the Couch with Freud and Kuhn," *Metascience*, DOI 10.1007/s11016-017-0253-3, and my *The Politics of Paradigms* (Albany: SUNY Press, 2019), ch. 8.

13. For an introduction to these political dimensions of Kuhn's theorizing, see my essay "The Paranoid Style in American History of Science," *Theoria* 27/3, no. 75 (Sept. 2012): 323–42. For a longer, partly biographical treatment, see my *The Politics of Paradigms*.

14. Ludwik Fleck, *The Genesis and Development of a Scientific Fact* (Chicago: University of Chicago Press, 1979). For a useful account of Kuhn's debts to Fleck, see Hans Joachim Dahms, "Thomas Kuhn's *Structure*: An 'Exemplary Document of the Cold War Era'?" in *Science Studies During the Cold War and Beyond*, ed. E. Aronova and S. Turchetti (New York: Palgrave, 2016, 103–25), esp. 112–15.

15. Kuhn to Ralph Lowell, March 19, 1950, TSK-MIT, box 3, folder 10. Kuhn to William H. Lawrence, April 13, 1950, Lowell Institute Records, Massachusetts Historical Society, carton 10.

16. This notion is central to a sophomore-year essay Kuhn wrote titled "The Metaphysical Possibilities of Physics" (TSK-MIT, box 1, folder 3), in which the word "infinite" in this and related senses appears repeatedly.

17. Thus Kuhn speaks of the mutilated chessboard puzzle, which he mentions at the end of Lecture V and the beginning of Lecture VI, as a paradigm for understanding methodological shortcuts that scientists employ to clarify and solve conceptual puzzles. Kuhn mentions James's "Bloomin, buzzin confusion" in Lecture VI, "Coherence and Scientific Vision."

18. *Structure*, 1.

19. On Frank and Kuhn see my *Politics of Paradigms*, ch. 10, and Frank's posthumously published book *The Humanistic Background of Science*, ed. George A. Reisch and Adam Tamas Tuboly (Albany: SUNY Press, 2021), esp. the editors' introduction, §3.3.

20. Quine had written, "The totality of our so-called knowledge or beliefs . . . is a man-made fabric, which impinges on experience only along the edges." Additional evidence for the influence of "Two Dogmas of Empiricism" is Kuhn's remark in an outline for the lecture series: "Analytic and synthetic not clearly distinct," he noted for Lecture VII ("Outline for Lowell Lectures, Spring 1951, TSK 3-30-1950," TSK-MIT, box 3, folder 10, p. 8). W.V.O. Quine, "Truth By Convention," in *Philosophical Essays for A.N. Whitehead*, ed. O.H. Lee (New York: Longmans, 1936), 90–124; "On What There Is," *Review of Metaphysics* 2(5) (September 1948): 21–38; "Two Dogmas of Empiricism," *Philosophical Review* 60(1) (1951): 20–43, 39.

21. *On Understanding Science* (op. cit.) served as a textbook while Conant and his instructors developed additional case histories that were later published as individual pamphlets and as a two-volume set titled *Harvard Case Histories in Experimental Science* (Cambridge, MA: Harvard University Press, 1957).

22. Alexandre Koyré, *Études Galiléennes* (Paris: Hermann, 1939).

23. On Kuhn's use of Piaget and other psychologists when writing *Structure*, see David Kaiser, "Thomas Kuhn and the Psychology of Scientific Revolutions," in *Kuhn's Structure at Fifty*, op. cit., 71–95. See also Kuhn's essay, "A Function for Thought Experiments," in *The Essential Tension* (Chicago: University of Chicago Press, 1977), 240–65.

24. Stuart Chase, *The Tyranny of Words* (New York: Harcourt Brace, 1938). Benjamin Lee Whorf, *Language, Thought, and Reality*, ed. John B. Carroll (Cambridge, MA: Technology Press of Massachusetts Institute of Technology, 1956).

25. Studies of Kuhn's "linguistic turn" include Stefano Gattei, *Thomas Kuhn's 'Linguistic Turn' and the Legacy of Logical Empiricism* (Farnham, UK: Ashgate, 2008); Amani Albeda, "A Gadamerian Critique of Kuhn's Linguistic Turn: Incommensurability Revisited," *International Studies in the Philosophy of Science* 20(3) (October 2006): 323–45; Alexander Bird, *Thomas Kuhn* (Princeton: Princeton University Press, 2000), ch. 5; K. Brad Wray, *Kuhn's Evolutionary Social Epistemology* (Cambridge: Cambridge University Press, 2011), 24–29.

26. See, for example, Steve Fuller, *Kuhn versus Popper: The Struggle for the Soul of Science* (New York: Columbia University Press, 2004).

27. *Structure*, 1.

28. Kuhn to Conant, June 29, 1961, TSK-MIT, box 25, folder 53. A.J. Ayer, *Language, Truth and Logic* (New York: Dover, 1936), ch. 3; *Structure*, 126, 98–103. A classic charge of Kuhnian irrationalism is Imre Lakatos's invocation of "mob psychology." See Imre Lakatos, "Falsification and the Methodology of Scientific Research Programmes," in I. Lakatos and A. Musgrave, eds., *Criticism and the Growth of Knowledge* (Cambridge: Cambridge University Press, 1970), 91–195, 178. On "conversion experiences" see, e.g., *Structure* 150, 151, and 203 in Kuhn's (1969) postscript. On "faith," see *Structure*, 157.

29. Carnap himself had proposed getting together with Kuhn to discuss common interests. See my essay "Did Kuhn Kill Logical Empiricism?" *Philosophy of Science* 58(2) (June 1991): 264–77. See also Gurol Irzik and Teo Grünberg, "Carnap and Kuhn: Arch Enemies or Close Allies?" *British Journal for the Philosophy of Science* 46(3) (Sept. 1995): 285–307; and Kuhn's essay "Afterwords," in *The Road Since Structure*, 216–52, 227. In a personal letter (Kuhn to George Reisch, August 19, 1993), Kuhn remarked about Carnap that "he and I might usefully talk" and expressed "intense

embarrassment" for having earlier doubted the sincerity of Carnap's admiration for *Structure*.

30. See, e.g., Carnap's two-part essay "Testability and Meaning," *Philosophy of Science* 3 (1936): 419–71; 4 (1937): 1–40; Otto Neurath, "Unified Science as Encyclopedic Integration," *International Encyclopedia of Unified Science*, v. 1, no. 1, (Chicago: University of Chicago Press, 1937), 1–27, esp. 23, 26. For an overview of Neurath's conception of unified science, see my "Planning Science: Otto Neurath and the *International Encyclopedia of Unified Science*," *British Journal for the History of Science* 27 (1994): 153–75. On Neurath's *index*, see my "Epistemologist, Economist…and Censor? On Otto Neurath's Infamous *Index Verborum Prohibitorum*," *Perspectives on Science* 5(3) (1997): 452–80.

31. Otto Neurath, "Foundations of the Social Sciences," *International Encyclopedia of Unified Science*, v. 2, n. 1 (Chicago: University of Chicago Press, 1944), 18, 6. *Ballungen* and their roles in Neurath's theorizing are explored in *Otto Neurath: Philosophy Between Science and Politics*, by Nancy Cartwright, Jordi Cat, Hasok Chang, and Thomas E. Uebel (Cambridge: Cambridge University Press, 1996). See, e.g., 80–81, 154–55. As early as 1916, in his historical analysis of theories of light and optics, Neurath emphasized the ineliminable "blurred margins" around and between scientific concepts. See "On the Classification of Systems of Hypotheses," in his *Philosophical Papers 1913–1946*, ed. and trans. R.S. Cohen and M. Neurath (Dordrecht: Reidel, 1983), 13–31.

32. See Verein Ernst Mach, *Wissenschaftliche Weltauffassung. Der Wiener Kreis* (Vienna: Wolf, 1929), translated as, "The Scientific World-Conception. The Vienna Circle" in F. Stadler and T. Uebel, eds., *Wissenschaftliche Weltauffassung. Der Wiener Kreis* (Vienna: Springer, 2012), 75–115. In a notebook of 1949, Kuhn listed the two monographs as "read in toto" (TSK, MIT box 1, folder 7). By the time Kuhn wrote *Structure*, he had also read Ernest Nagel's encyclopedia pamphlet, "Principles of the Theory of Probability," *International Encyclopedia of Unified Science*, v. 1, n. 6 (Chicago: University of Chicago Press, 1955 [orig. 1939]), which he cites in the chapter "The Resolution of Revolutions"; and Carl Hempel's "Fundamentals of Concept Formation in Empirical Science," *International Encyclopedia of Unified Science*, v. 2, n. 7 (Chicago: University of Chicago Press, 1955 [orig. 1952]), which he mentioned reading in personal correspondence (Kuhn to George Reisch, August 19, 1993). Another relevant text Kuhn had read by 1949 was A.J. Ayer's *Language Truth and Logic*, op. cit.

33. Woodger demonstrates how a simple biological theory may be exhaustively specified in logical symbols and statements, but does not call for his formalization to replace ordinary language. His endorsement of formalization was also pragmatic and experimental: "Today it is futile to dogmatize about the possibilities of applied logic," he wrote. "What it can do can only be discovered by trial." Joseph Woodger, "Techniques of Theory Construction," *International Encyclopedia of Unified Science*, v. 2, n. 5 (Chicago: University of Chicago Press, 1970 [orig. 1939]), 449–531, see quote on 523; see also 454, 522. Bloomfield's monograph speaks of ordinary language as an "ordering and formalization" of the "flowing phenomena of the universe" and regards formal analysis and reconstruction as a tool to explore the "vagueness at the borders" of words and concepts. Even though a "system of formal logic" may be called a "language" for science, he emphasized, logical systems are themselves empty without interpretation through a natural language. Leonard Bloomfield, "Linguistic Aspects of Science," *International Encyclopedia of Unified Science*, v. 1, n. 4 (Chicago: University of Chicago Press, 1955 [orig. 1939]), 215–339; see 239, 249, 262–64.

34. Alfred Korzybski, an influential champion of popular semantics, proposed to eliminate Aristotelian logical foundations from everyday speech, including use of "is": "Once we abolish in our language the always false to fact 'is' of identity, we automatically stop identifying different orders of abstractions." See his *Science and Sanity: An Introduction to Non-Aristotelian Systems and General Semantics* (Lakeville, CT: The International Non-Aristotelian Library Publishing Company, 4th ed., 1958), 474.

35. John Dewey, "Unity of Science as a Social Problem," in "Encyclopedia and Unified Science," *International Encyclopedia of Unified Science*, v. 1, n. 1 (Chicago: University of Chicago Press, 1937) 29–38, 34. Horace Kallen, "The Meanings of 'Unity' Among the Sciences," *Educational Administration and Supervision* 26(2) (1940): 81–97, 91.

36. For more on these debates and their proximity to Kuhn, see my essay "What a Difference a Decade Makes: The Planning Debates and the Fate of the Unity of Science Movement," in Jordi Cat and Adam Tamas Tuboly, eds., *Neurath Reconsidered: New Sources and Perspectives* (Springer Verlag, 2019), 385–411.

37. Conant to John Bower, Sept. 12, 1952, Harvard University Archives (UAI5.168, Box 467). Courtesy of the Harvard University Archives.

38. James's phrase appears in *Structure*, 113. On revolutions as "conversion experiences," see, e.g., 150, 151, and Kuhn's (1969) postscript, 203. Kuhn's notecard on James's *The Varieties of Religious Experience* is dated 12/30/43 (TSK-MIT, box 8). I thank Juan V. Mayoral for sharing this document with me.

39. *Road Since Structure*, 289.

40. For example, in his eagerness to inform his audience that the story of Galileo dropping balls of different weights from the Tower of Pisa is apocryphal, Kuhn dismisses Galileo's observation that a heavy ball reaches the ground sooner than a lighter ball "by a long space" on the grounds that this report is incompatible with "the laws of motion." In the next paragraph, however, Kuhn explains that because of air resistance "the heavier one gradually gets ahead and hits the ground first," as Galileo had reported.

41. Edward Weeks, *The Lowells and Their Institute*, 162. When asked whether he recalled attending or hearing about Kuhn's Lowell lectures, Kuhn's colleague Gerald Holton replied that he did not and pointed out that scholars from Harvard and MIT were unlikely to have attended given Kuhn's relative obscurity at the time and because then, as now, "oodles of lectures" were available to scholars every day in Cambridge (personal communication, June 6, 2018).

42. This flyer is reproduced here on pp. viii–ix.

43. These include, for example, Albertus Magnus and the "Parisian nominalists in the century which followed," physics laboratory demonstrations such as Newton's rings, and the emergence of logicism in the philosophy of mathematics: "As many of you realize," Kuhn remarked early in Lecture VII, "the theorems of mathematics can be derived from the more general principles of modern logic." After assuming this level of logical sophistication for his audience, Kuhn explained at length mere rudiments of propositional logic. These kinds of difficulties appear to have frustrated other aspects of Kuhn's early career. His first book, *The Copernican Revolution* of 1957 more effectively reached academic and popular audiences; but that success reportedly worked against his bid to obtain tenure at Harvard. See James A. Marcum, *Thomas Kuhn's Revolution: An Historical Philosophy of Science* (New York: Continuum, 2005), 13–14.

44. *Road Since Structure*, 280, 289.

45. See Kalman Seigel, "College Freedoms Being Stifled By Student's Fear of Red Label," *New York Times*, May 10, 1951. For a detailed survey of how American scholars were affected by McCarthyism, see Ellen Schrecker's *No Ivory Tower: McCarthyism and the Universities* (New York: Oxford, 1986).

46. Kuhn's alarm over this publicity is evident in his correspondence with Ralph Lowell. He noted his "acute distress" in a letter of Feb. 20, 1951 (TSK-MIT, box 3, folder 10). For more on this episode and related issues, see my *Politics of Paradigms*, ch. 9.

47. *Road Since Structure*, 289.

48. On Kuhn's proposal to replace "paradigm" by "disciplinary matrix," see the postscript to *Structure*'s second and subsequent editions (*Structure*, 173–208, esp. 181). In a prefatory note to the second edition of his book on Planck, Kuhn points out that the book's first edition is "scrupulously silent" about its relationship to *Structure*'s theory of science (*Black-Body Theory and the Quantum Discontinuity, 1894–1912*, 2nd ed., Chicago: University of Chicago Press, 1987), xv. Kuhn discusses this relationship in that edition's afterword (pp. 349–70).

49. In Lecture V, Kuhn raises the specter of dogmatism by placing aside the popular view of scientists as "open-minded" to suggest that young scientists are instead empty-minded and informed by education ("I contend then that emptiness rather than openness characterizes the youthful mind"). Kuhn would later emphatically dismiss the ideal of open-mindedness in in his essay "The Function of Dogma in Scientific Research," in *Scientific Change*, ed. A.C. Crombie (New York: Basic Books, 1963), 347–69. On Kuhn's more individualistic approach in his Lowell lectures, see Juan V. Mayoral, "Five Decades of *Structure*: A Retrospective View," *Theoria* 27/3, no. 75 (Sept. 2012): 261–80, esp. 267 and 271; see also K. Brad Wray, "The Influence of James B. Conant on Kuhn's *Structure of Scientific Revolutions*," *HOPOS* 6 (Spring 2016): 1–23.

50. Kuhn mentioned "the completion of my script" in his correspondence with the Lowell Institute (Kuhn to William H. Lawrence, Nov. 8, 1950, Lowell Institute Records, Massachusetts Historical Society).

51. *Road Since Structure*, 284.

52. This edition of *The Quest for Physical Theory* could not exist without the generous assistance of the archival staff at MIT Libraries Department of Distinctive Collections and copyright permissions arranged by Katie Zimmerman, Director of Copyright Strategy at MIT Libraries. For help in securing permissions I thank Sarah Kuhn, and for help in tracking down a handful of missing photocopied pages from the MIT archives I thank David Kaiser and Lauren Kapsalakis. For permission to quote from correspondence held in the Lowell Institute Records, I thank Hannah Elder at the Massachusetts Historical Society and for permission to quote correspondence between Kuhn and Conant held in the Harvard University Archives I thank Juliana Kuipers. Last I thank several colleagues who fielded queries related to these lectures and offered helpful comments on my introduction, including K. Brad Wray, Gerald Holton, Robert J. Richards, Adam Tamas Tuboly, Thomas Uebel, Juan V. Mayoral, and an anonymous reader.

COURSE TICKET

This ticket entitles the bearer to any seat not reserved for the use of the Trustee or the Lecturer, in the Lecture Hall of the Boston Public Library, Boylston Street, Boston.

NO SEATS RESERVED AFTER 7.55

LOWELL INSTITUTE

EIGHT LECTURES

on

THE QUEST FOR PHYSICAL THEORY, PROBLEMS IN THE METHODOLOGY OF SCIENTIFIC RESEARCH

by

THOMAS S. KUHN, A.M., Ph.D.

Tuesday and Friday Evenings at 8 o'clock

Beginning Friday, Mar. 2, 1951 and omitting Mar. 23

Hall doors opened at 7.30; CLOSED at 8.00 o'clock AND NO ONE ADMITTED AFTER THE LECTURE BEGINS

Lecture I

Introduction: Textbook Science and Creative Science

Mr. Lowell, Professor Lawrence, ladies and gentlemen: good evening.[1] Before turning to this evening's topic let me take a moment to express my appreciation of the Lowell Institute's gratifying invitation to participate in this famous lecture series. My pleasure in accepting Mr. Lowell's kind offer is made the greater by my consciousness of the intellectual debt I share with many of my contemporaries to the eminent scholars who have preceded me on this platform. The tradition which they have established I cannot hope to emulate; but perhaps I may repay a small portion of my indebtedness. With that objective let me proceed to my topic.

Shortly before the beginning of this century, the British statistician and philosopher of science Karl Pearson provided a classic statement of a point of view which is implicit today in much that is said and written about the nature of science and of scientific knowledge. Briefly stated, Pearson's position was that the validity and utility of scientific knowledge derives not from its subject matter, but from the universal method employed in gaining it. And the basis of this method, from which both its power and its universality proceeded, lay for Pearson in the dispassionate observation and classification of the objective facts of the natural world. "The scientific man," he said,

> has above all things to strive at self-elimination in his judgments, to provide an argument which is as true for each individual mind as it is for his own. The classification of facts, the recognition of their sequence and relative significance is the function of science, and the habit of forming a

judgement upon these facts, unbiased by personal feelings, is characteristic of what may be termed the scientific frame of mind.²

For Pearson, the scientist was "the man who has accustomed himself to marshal fact, to examine their complex mutual relations and predict upon the result of the examination their inevitable sequences ... which we term natural laws ...." "Such a man," he said, "... will scarcely be content with merely superficial statement, with vague appeal to the imagination, to the emotions, to individual prejudice; he will demand a high standard of reasoning, [and] a clear insight into facts and their results.³

This description of the scientist as a man who collects and classifies facts so that he may proceed to the prediction of inevitable sequences did not originate with Karl Pearson or in our own century. On the contrary, it is as old as modern science itself. One need only remember Francis Bacon's proclamation of the New Method at the beginning of the seventeenth century:

> The true method of experience [. . .] first lights the candle and then by means of the candle shows the way, commencing as it does with experience duly ordered and digested [. . .] and from it deducing axioms, and from established axioms again new experiments, even as it was not without order and method that the divine word operated on the created man.⁴

Similar sentiments have been expressed repeatedly since Bacon's time, both by scientists themselves and by those philosophers, from John Locke to John Stuart Mill, who have felt impelled to account for the progress of scientific knowledge. Pearson's statement was simply the clearest and his claim of universality the most ambitious, so that, however much one dissents from the detail with which he develops and documents his thesis, one must recognize in it an expression of the essence of the empiricist methodological tradition which has dominated the discussion of science since Bacon's day.

Twentieth-century methodology has achieved a subtlety of analysis neither dreamed of by Bacon nor realized by Pearson. But the basic premise of empiricism, that the scientist, or at least the ideal scientist, proceeds from objective experimental facts or meter readings to the unique laws which govern them, still underlies the most sophisticated methodological discussion. And this same conception of the godlike objectivity of science infects our very language. To be *scientific* in an argument or an analysis is to be something worthy of praise; nothing annoys us more to be told that our pet theory is "unscientific," for this we take to mean biased or based on personal prejudice. Even though our contemporary disillusionment has deprived us of our image of the scientist as a messiah, we preserve our symbol of the scientist as the man in the highly starched, gleaming white coat who, in the laboratory as in the dentifrice ad, abandons all prejudice so that he may proceed first to a dispassionate analysis of all the facts and then to the formulation of the immutable law which governs them.[5]

Now I think that this picture of the scientist, and the correlated description of the method by which the scientist reaches his conclusions, is altogether wrong. In saying this, I do not mean simply to point out that of course scientists are human and that since "to err is human," they can seldom measure up to the rigid standards set by their methodological canons. Such individual human failures are of course both inevitable and trivial. Their existence cannot detract from Pearson's description, if that description is taken to be an ideal toward which the individual scientist must strive if he is to succeed in his research.

I mean rather to deny even the normative validity of Pearson's canons. And, as an alternative, I should like to suggest that the impartial, dispassionate observation of nature is impossible, that there are no "pure facts" from which alone valid theories can be derived, and that the effort toward "self elimination" which Pearson proposes as the scientist's goal would, in practice, result in the abolition of productive research. In short, I believe that the elements which, on Pearson's description, can only be called prejudice and preconception are inextricably woven into the pattern of scientific research, and that any attempt to eliminate them would inevitably deprive research of its fruitfulness.

Much of this lecture and of the seven which are to follow will be devoted to a description of the evidence which has led me to so radical a conclusion, but I hope that these lectures may also serve a more constructive function. For I was trained as a physicist; I have done research; and I am committed to the belief that scientific knowledge is good knowledge, that it is useful knowledge, and above all that it is cumulative knowledge. I believe that the work of scientists has resulted in an increasingly detailed and an increasingly far-reaching understanding of the operations of nature, and that this progress can be objectively described so that a seventeenth-, eighteenth-, or nineteenth-century scientist would, without reluctance, admit that science had proceeded a long way since his own day.

Therefore, in denying the possibility and the desirability of dispassionate neutrality on the part of the individual scientist, I do not at all mean to deny an objective sort of validity to the products of the scientific profession, and it will therefore be a second and more basic task of these lectures to provide an alternative description of the method employed by the scientist, and to show how science conceived as a body of cumulative knowledge can proceed from this alternate description. In the broadest and most fundamental sense, the objective of these lectures is then to provide a preliminary description of scientific activity and to discover the relationship of this, the activity of the working scientist, to the products of his profession, to science as a body of human knowledge.

For this purpose I urge the utility of separating sharply in our minds two distinct meanings of the word "science." In the first of these science is conceived as an activity, as the thing which the scientist does. In its other meaning science is knowledge, a body of laws and of techniques assembled in texts and transmitted from one scientific generation to another. These are the two meanings distinguished in the title of this evening's lecture as Textbook Science and Creative Science.[6] And it is the search for the relation between these two which will constitute the primary objective of our study. To the extent that we are successful we may hope to learn something of the nature of scientific knowledge.

* * *

If any of you happens to have followed closely the advance notices of this series of lectures, you may have remarked that the topic just described bears very little relation to one announced in some of the flyers prepared by the Lowell Institute's copy writer. That topic was, I believe, described under a banner head reading: "What Are the Problems of Scientific Research Today?"[7]

I can scarcely imagine a more fascinating question; I should gladly attend a series of lectures devoted to it. *Except* that I doubt whether any serious student of science or scientific method would consider himself equipped to address such a subject. Therefore, with apologies for any confusion that the misrepresentation may have created, I should like to announce that I do not intend to deal with any of the problems raised by that question at any point in this series of lectures.

My own topic is quite ambitious enough, and the goal I have set is a very distant one. Actually the description with which I have thus far provided you is a description of a continuing research program of a rather unusual, though not unique, sort, and these lectures are intended to serve as an introduction, a Prolegomenon, to this new field of research rather than as a report on the outcome of a completed study. And even so the program is too broad, so I should like immediately to introduce two further restrictions of the subject matter with which I intend to deal.

In the first place, on the grounds of personal competence, I shall restrict my attention to that part of science generally described as physical, that is, to those sciences which deal with the workings of inanimate nature, particularly physics, chemistry, and astronomy. I should of course like to suppose that the remarks I will make apply equally well to other sciences. I should be less concerned with the field of study if I did not suspect that it had far reaching implications. But my own conclusions are drawn from a study of the physical sciences, and the judgment of their applicability to other fields must finally be left to the man working in those fields.

Second, even within the body of physical sciences, our attention will be directed toward only that portion of scientific research which eventuates in new conceptions about the material world. We shall be concerned with the sort of research the led to the Newtonian laws of motion, not with the manner in which these laws were applied in building new machines or instruments. We shall be concerned with the work of such men as Boyle and Dalton, insofar as this led to a new understanding and a new set of laws governing the formation of chemical compounds, but we shall not be concerned with the manner in which these laws, once arrived at and confirmed, were applied to the production of dyes, explosives, or plastics. Of course the importance of science lies at least as much in its application as in the fundamental insights it provides into the workings of nature, but if we are to reach any conclusions our concern must be restricted, and I have chosen to direct your attention and mine to the conceptual rather than the tangible aspects of scientific progress.

Finally I should like to introduce at the very start a further and very important qualification of our subject matter, a qualification which concerns the extent to which we shall be dealing with the theories of contemporary science. The sorts of methodological conclusions I wish to draw are of primary significance only to the extent that they apply to contemporary physical science. Nevertheless, in drawing these conclusions and in supporting them, I shall make use of historical material drawn almost entirely from scientific developments before the twentieth century. I believe that the historical unity of science, or more accurately the historical unity of scientists, permits the picture of science which we shall derive in this matter to be applied without significant alteration to contemporary science, and I trust that before we are through you too will discover that this must be the case. Illustrative material drawn from contemporary science would undoubtedly provide more complete evidence, but there are two considerations which override my desire to complete the case for you.

Modern physical science is a highly abstract and highly technical field relatively unfamiliar to most of you. To attempt to discover the manner in which scientists arrived at the conclusions contained in the theories of quantum mechanics or relativity would require our first determining what these conclusions are, and since the historical or genetic approach is neither the clearest nor the briefest manner of presenting scientific conclusions, we would almost certainly lose our way.

But there is another far more significant reason for refraining from a close examination of contemporary science, a reason which is rooted in and which displays the nature of our objective. For we are concerned to explore the way by which scientific theories come into existence and to discern the relationships between particular mental and experimental processes and the finished conceptual schemes to which they give rise. This is particularly hard to achieve in dealing with the science in which we happen to believe. For the theory in which we believe is necessarily and uniquely characterized by the apparent inevitability of its relation to the facts from which it arose. Our belief itself represents a commitment to the double position that only this theory will account for the facts which we know *and* that this theory will account for all the relevant facts. We may admit that in the future there will be other facts and other theories, but for the moment we cannot conceive these, so the appearance of the inevitable connection remains.

In dealing with older scientific theories we gain a great advantage. We are committed neither to these theories nor to the mode of thought which gave them birth, and so we may at least hope to find answers to such questions as:

"Why did this set of experimental findings lead to this theory, rather than to the alternate one we hold today?" or "Why didn't this 'fact' appear relevant to a test of the validity of that theory?" It is in answering such questions that we shall trace the progress of the mind in its pursuit of scientific theories, and it is in this pursuit that whatever may properly be called scientific method is to be found. Accordingly, you will hear very little about twentieth-century physics during the course of these lectures.

This discussion leads us back to the point from which we started. For I should now like to suggest to you that the dominant empiricist methodological tradition, whose principal tenets we described earlier, gains its eternal plausibility because it is drawn from an examination of contemporary science, or the science which its author believes. Put more precisely, I believe that Pearson's methodology and many of the others which resemble it are drawn from a study of the finished products of scientific research, that is, from the study of textbook science. For if we turn our attention for the moment from the procedures of the working scientist to the form in which the final products of the scientific profession are delivered and transmitted, we find again the insuperable division between the immutable law and the objective dispassionate experiment which confirms the law, and it is just this division which characterizes all empiricist methodologies like Pearson's.

In a textbook we are presented with a law governing the behavior of certain aspects abstracted from nature, and we are made familiar with certain of the rules of logic and of mathematics by which we may deduce particular consequences of the law. These consequences may be tested by a set of operations also prescribed in the text to see whether or not they in fact correspond with the operation of nature. Finally, we are given, usually in tabular or graphical form, the result of such a series of manual operations and a corresponding series of deductions from the law, and we are asked to judge whether the agreement between the two is sufficiently close to justify our holding to the law. If it is a good textbook our judgment is always in the affirmative.

This is the way we write up the products of our research, and this is the way we teach scientific theory to our students. We report to them that Galileo stated the law that the distance traversed by a body falling from rest is equal to one-half the acceleration multiplied by the square of the time during which the body has fallen and that this holds for all bodies regardless of their weight. More briefly, we tell our students that $s = 1/2\ at^2$, the famous law which causes so much trouble in elementary physics courses. If our students refuse to take our word for this, we build or order from the nearest physical supply laboratory an apparatus which will demonstrate it quite convincingly. One excellent piece of equipment which I have used for this purpose employs an electromagnet which releases an iron bob at a prescribed instant, and an electrified timer which activates a spark gap and thus causes the falling bob to mark its position on a specially chemically treated paper tape every tenth of a second during its fall. Galileo, of course, had no such equipment. But that is not relevant to our attempt to validate his laws.

If our students are still skeptical and say that this law is fine but that it holds only for heavy metal bodies and would not hold for a scrap of paper or a feather, we remind them that air resistance enters into these phenomena, and we build, again with considerable trouble and expense, an elaborate air pump, in the vacuum provided by which we can drop a lead weight and a feather. As predicted, these two then fall with the same speed.[8]

This is perfect textbook or pedagogic procedure: we give the student a law—$s = 1/2\ at^2$; we show him how to measure s with a meter stick or a vernier; we tell him, or we show him how to determine, the value of the acceleration a; and we teach him, or assume that he knows, the laws of multiplication which will enable him to compute the product of one-half times a time t squared, t of course being the time. We then let him work out the values of s, the distance fallen, for a variety of values of t, the time, and then demonstrate with our prepared apparatus that in fact the body actually falls according to the results which he has computed for himself. At this juncture, if he is not already asleep, he believes us and goes home happy.

As a student, setting out to learn the presently accepted laws of physics, he is well advised to credit our demonstration. For excluding the weight of our authority which is probably and unfortunately the true source of his belief, we have no other means to produce conviction. If our student chooses to doubt the validity of our tests, and he may have good reasons for doing so, we can go no further; as proofs for accepted laws the scientist can provide only observation and experiment—there are no ulterior tests. I doubt that there is any generalization about science of such universal validity as the one which states that any scientific law or theory *which has borne the stamp of approval of the profession for a number of years* has among its consequences predictions which can be thus exposed to experimental test. Further, these experimental tests are such that they can be performed by any man willing to assemble the necessary equipment.⁹

I believe that the validity of this generalization about textbook science is the true source of the empiricist methodology which we examined at the beginning of this lecture. And I would like to suggest that the error made by the empiricist lies in his plausible and implicit assumption that what is true of science in one of its meanings must be true of it in another. Because textbook science, which is the science that we know, proceeds from the statement of laws to the description of objective experimental tests of these laws, we are vulnerable to the belief that the creative science which lies behind the textbook must have pursued the reverse route, must have proceeded from the objective experiment to the law. We assume that the structure of knowledge in the textbook, the structure which we give to scientific knowledge for its transmission and preservation, provides a substantial clue to the nature of the creative process by which we gained that knowledge. And it is from this assumption that I should like to dissent.

There are a number of quite distinct reasons for this dissent: reasons drawn from logic, from the study of language, and from psychology. We shall touch upon a few of those this evening, and upon more of them later in these lectures. But at the moment I should like to concentrate particularly upon what might be called the historical reasons, for it is primarily by an examination of scientific practice as exemplified in the history of science that we may hope to get a more nearly correct notion of the procedures of creative science. But in turning to history as a source of data for the reconstruction of scientific method, we must be particularly careful. For the history of science is a relatively recent field of scholarly research: it has scarcely yet been awarded the mantle of academic respectability. And much of what currently passes as history of science belongs more properly to mythology—a mythology which has been created by reading history backward, by assuming that the man who first enunciated a particular law must have derived it from much the same evidence which we should now employ in documenting it.[10]

It will both reinforce my point and illuminate the problems with which we must deal if we now examine an example of such a fable. May I therefore introduce you once again to a story with which I am sure you are all familiar, to the story of Galileo whose experimental study of the motion of heavy bodies is said to have broken the stranglehold of Aristotelian dogmatism and to have paved the way for the great Newtonian generalization of the motions of terrestrial and celestial dynamics. It is particularly suitable to begin our study with the work of so great a scientist.

The usual version of Galileo's accomplishment is one which you all know. It was Galileo who, by dropping two bodies of different weights from the Leaning Tower of Pisa, first showed the world that two bodies of different weights, if dropped simultaneously from a high place, reach the ground together. It was again Galileo who first undertook a careful investigation of the way in which bodies actually do fall—an investigation in which by a stroke of the greatest genius he made use of the inclined plane—and it was Galileo who thus discovered the law which we've already discussed, the law $s = 1/2\, at^2$. Again it was Galileo who, by experiment, determined the important properties of the pendulum, and who thus made possible the development of the first accurate timekeeper, the pendulum clock, and the consequent development of accurate astronomical and navigational techniques. Galileo has thus been referred to again and again as the father of experimental science, for according to this version of his work it was he who for the first time decided that the true question for science to examine was *how* physical bodies actually behave, and accordingly it was he who first went out to make actual measurements of their behavior.

I do not want to detract from the fame so deservedly associated with the name of Galileo, but there is something very wrong with this plausible and generally credited account of his method of research. In the first place, it is very bad history. We possess, for example, a number of accounts of Galileo's experiment from the Leaning Tower of Pisa, accounts which describe in great detail the size, shape, and material of the bodies which he dropped and which provide as well the names and number of the disciples who accompanied him on his pilgrimage to the altar of truth. But most of these accounts were written in the last hundred years, and we know them to be fabrications.

Galileo, we now believe, did not perform this experiment at all. Although his works are filled with references to experiments (some of which he could not possibly have performed), there is no reference to this one either in his work or in that of his friends and contemporaries at Pisa. Our only seventeenth-century source for the story is a brief account provided sixty years after the presumptive date of the experiment by Galileo's disciple Viviani, who himself was born thirty years after the close of Galileo's residence in

Pisa. And even though we should like to credit this lonely and inadequate scrap of evidence, we cannot. For during his stay in Pisa, at the time when the experiment is supposed to have been performed, Galileo wrote a treatise *On Motion* in which he stated views about the behavior of falling bodies which are incompatible with his having performed the experiment.

This passage deserves further attention. In it Galileo discusses two falling bodies, one of lead and one of wood, released simultaneously from a high tower. Of these he says:

> Experience shows . . . [that] in the beginning of its motion the wood is carried more rapidly than the lead; but a little later the motion of the lead is so accelerated that it leaves the wood behind; and if they are let go from a high tower, precedes it by a long space; and I have often made a test of this.[11]

Here is a passage which could refer to an experiment from the Leaning Tower, but then how can we interpret its outcome? The result Galileo quotes is not compatible with his famous law: he says the heavy body gets way ahead. And unless the laws of motion have changed since Galileo's day, this was not the case. So presumably Galileo is here referring to an experiment which he never performed, or which, if by any chance it was performed, did not yield results compatible with his law.

The scientist ought not, I think, be very much surprised at these historical discoveries, for he should know that an experiment like that from the Leaning Tower of Pisa is a peculiarly bad one with which to demonstrate the truth of Galileo's law. If you drop two bodies of roughly the same size and shape but of quite different weights, from a low height, say something less than six feet, the two will hit the ground at about the same time. They will, in fact, strike so nearly simultaneously that it will be impossible without a photo-cell and a camera to tell which struck first. But this is a pretty crude experiment. The time of fall is so short that it does not allow differences in the speeds of the bodies to become apparent. And besides, these differences, if they exist, are much smaller than the differences due to the inevitable human failure to release the bodies quite simultaneously. Sound conclusions can only be drawn from the results of an experiment involving a longer fall. But if you carry the two bodies to a great height and drop them from, say, the Leaning Tower of Pisa, you discover that although the two bodies seem to fall together at the start, the heavier one gradually gets ahead and hits the ground first. When, in 1641, a Pisan correspondent of Galileo's actually tried the experiment from the Leaning Tower, he got just this result: the lead always beat the wood to the ground.

Today we know that the apparent difference between the results of the experiment at great heights and that at small heights is caused by the resistance of the air, which, *ceteris paribus*, acts more strongly upon the lighter body; as historians we may even suspect that it is this difference between the results of the two experiments which was responsible for the confusion already noted in the early work of Galileo (and in that of his sixteenth-century predecessors) about the differences in the behavior of a falling body at the beginning and the end of its motion. But these modern subtleties do not touch the real problem presented by the historical investigation of the story of Galileo and the Leaning Tower.

For neither the experiment from the Tower nor the experiment on the ground could show more than that Aristotle was wrong—that the time required to reach the ground did not vary in proportion to the weight of the body. And this fact was already known: it was probably known in antiquity; it was certainly known to two important commentators on Aristotle who worked in the sixth century of the Christian era; and it was widely circulated in Europe from at least the thirteenth century on. This was not Galileo's original contribution—he almost certainly learned it in school. His own contribution was the statement of a new law, the law that the time of fall does not depend on the weight. And this law he did not get by an experiment from the Leaning Tower. For in the first place, he almost certainly performed no such experiment, and, secondly, if he had performed it carefully, it would not have confirmed his law.

We meet a similar difficulty in examining Galileo's experiment with the inclined plane, the experiment with which he is supposed to have discovered that the distance through which a body falls or rolls is proportional to the square of the time consumed by rolling. The law is true, and the experiment was almost certainly performed by Galileo. But he knew the law before he performed the experiment. And if he had not known it, the experiment would not have provided it, for the equipment available to him was too crude. Father Mersenne, the French Franciscan friar whose correspondence played so large a part in spreading scientific knowledge through Europe in the seventeenth century, attempted to repeat Galileo's experiment using a carefully constructed duplicate of Galileo's equipment. His results diverged so far from those quoted by Galileo and from those which would have been predicted by Galileo's law that Mersenne was finally forced to "doubt whether Galileo had performed the experiment at all."[12] Mersenne's doubt is not shared by most present-day historians. Galileo probably did perform this experiment. But when he did so he did not confirm his law with the accuracy of which he tells us, an accuracy which, in his own words, was "such that the deviation never exceeded one-tenth of a pulse beat."[13]

The inclined plane can be made to yield Galileo's result. In fact we have recently built such an inclined plane at Harvard for use in demonstrations in the elementary science program. Our inclined plane, in contrast to Mersenne's and Galileo's, gives the "right" result; but its construction required several months of careful thought, the operation of the best modern machine tools, and the expenditure of something over five hundred dollars. Galileo would not recognize his equipment in our form, and he did not arrive, and could not have arrived, at his conclusion with his own.[14]

I wonder whether there is any need to continue the enumeration of examples of this sort. For we started this examination of the "fable of Galileo" not in order to correct misconceptions about the work of Galileo, but in order to document a conclusion and to raise a series of questions which will be of fundamental importance throughout these lectures. The conclusion is the one which I suggested at the start: the mythology of science shows a remarkable resemblance to textbook science;[15] that is, in the present case, the fable of Galileo displays its hero deriving his laws from exactly the experiments which we should now use to validate them.

But since I have already suggested that textbook science is the source of empiricist methodology, of the methodology of Pearson *et al.*, our conclusion leads us directly to a central question: can we salvage the empiricists' methodology by examining the true history of science? If Galileo did not get his laws from the experimental facts which we should use in validating them, are there some original experiments from which he did derive them? And on this point Galileo's own testimony is not without significance. In one passage he states that "in order to be able to demonstrate to my opponents the truths of my conclusions, I have been forced to demonstrate them by a variety of experiment, though to satisfy myself alone I have never felt it necessary to make many."[16] Or again, in a passage from the *Dialogues* which more clearly reveals the essence of Galileo's method, the author's spokesman declares to an Aristotelian opponent: "I, without having made an experiment, am certain that the effect will follow just as I have described it, for it is necessary that it do so. Furthermore, although you pretend the contrary, you too know that it could not happen in any other way. But I am such a good midwife of the mind that I will force you to confess the truth of what I say with all your heart."[17]

These evaluations with their dialectic tone and their clear statement of the supremacy of the intellect over experiment and manipulation are not, I think, typical of science in general. But they are entirely typical of Galileo and science has never ceased to use the method which he so profitably employed. What this method was, how Galileo actually arrived at his laws, is a question whose answer I shall reserve for the next lecture. But we are already in a position to recognize that whatever the method, it was not primarily based upon original experimental discovery. As we shall find, the experimental and observational basis of Galileo's laws had, with the exception of the observation of the pendulum, been available to all scientists concerned with the problem of motion since antiquity. And this strongly suggests that we will not be able to salvage empiricist methodology. Galileo, if a poor and frequently irresponsible experimenter, was one of the greatest scientists the world has ever known. Yet he was not, to revert to Pearson's phrase, a "man who has accustomed himself to marshal facts, to examine their complex mutual relations and predict upon the result of this examination their inevitable sequences." For there were no such "facts."[18][19]

So far I have attempted to indicate the sort of evidence with which I think the history of science can supply the methodologist. And this turns out to be a point requiring considerable proof, for it has not normally been conceded that history and methodology have any grounds of mutual reference. Certainly our case for the use of history is still far from complete. We have, this evening, seen history only in a negative, a destructive, role. My homicidal attack upon the fable of Galileo was not calculated to produce final or permanent conviction. I merely hope that you have found it suggestive.

In the three lectures that follow this evening's we shall deal in considerably more constructive detail with selected portions of the history of physical science. And this less-destructive examination of science will, I trust, provide material from which during the last four lectures we can begin to construct an alternative picture of creative science. But in this latter part of the series we shall utilize not only historical techniques but also those drawn from logical and psychological criticism. And it may well round out this evening's introductory remarks if I attempt to give a preliminary indication of the applicability of such nonhistorical techniques to our problem. Here once again I shall touch in brief and destructive terms upon a subject to which we will turn with more constructive detail later in the series.

The suggestion from which I should like to proceed is that we ought not be surprised that a naive empiricism stands up so badly when applied to the examination of creative research. For, as I indicated earlier, how could one set out to make the pure and dispassionate observations for which a naive empiricism calls? Doesn't one require—at the start and throughout the research—some guiding principles? Suppose, for example, that I am directed as a scientist to begin observations of this rostrum in order to make some scientific discovery about it. I am directed, we'll say, to look for the laws of the rostrum. The very existence of this directive raises a difficulty. For why should it be supposed that there are any such laws? Why should I examine the rostrum rather than the platform, the lighting circuit, or the people in this auditorium? Since in my lifetime I will not be able to study them all, the decision to examine the rostrum implies a judgment that there are laws to be discovered there and that these will be useful laws. And this judgment must have been made prior to the beginning of my examination.

Well, suppose that there is a basis for such a judgment. Suppose that there is reason to think that there will be new laws produced by my examination, laws which are not already in the body of science. What observations or experiments should I then make? I might start out by finding how much the rostrum weights. I bring in a scale and begin the usual determination of weight. And immediately another question arises: Will the normal weighing operation be adequate? Is it sufficient for my purpose to know the weight of the rostrum to the nearest pound, as though I were going to send it somewhere by Railway Express? Or will the discovery of a valid scientific law require the utmost of accuracy? Will I need to know the weight to within a few one hundred thousandths of a pound? If the latter is so, I shall have to design and construct a special sort of scale, and I may also require an air pump which will enable me to do the weighing in a vacuum. All this will take a great deal of time, trouble, and expense, and since my useful lifetime as a discoverer is limited, these can only be spared if they are likely to prove worthwhile. Once again a decision, a judgment is called for.

An even more serious problem obtains with respect to my next step, for I now wish to observe the dimensions of the rostrum. Again I must decide what accuracy, what instruments are required. But even with this decision made my problem is far from solved, for I don't know which dimension to take. Will the height, the length, and the breadth be sufficient for my purposes or must I measure the various diagonals. Can I describe the edges of this panel as being straight, or must I carefully measure the deviations of each edge from a mathematically straight line? The process of obtaining complete measurements of the size can never be completed. I must make a selection.

But perhaps I need not investigate the dimensions at all. Perhaps they are not relevant to the discovery of new scientific laws. Perhaps what I need is a microscopic examination of the grain structure of the material from which the rostrum is made; or, would I be better employing my time applying to my problem the tools of qualitative and quantitative chemical analysis? May I not require knowledge of the proportions in which the various chemical elements enter into the construction of the rostrum? Or ought I rather be measuring the strengths of the materials of which it is built? And any one of these choices raises the old problems of accuracy and of selection.

At this point I am struck by a truly horrifying thought. Perhaps none of these techniques of modern scientific measurement is the one required to yield the law of the rostrum. Perhaps a new sort of measurement is required in this case. And indeed science is continually proving the fruitfulness of heretofore unimagined techniques of observation. But then how shall I go about discovering these? It seems hopeless. And now I am so puzzled and confused by the attempt to regulate my research procedures in conformity with the canons of pure empiricism that I give up the attempt entirely. Instead I will take up cooking. There at least one has a cookbook.

But, of course, in science one has a cookbook, too. The choice of a problem, the selection of tools, the decision as to which aspect of the object should be abstracted for scientific consideration, all of these require evaluation and a judgment before beginning research. And as the investigation proceeds, similar decisions are called for: what variables must be controlled, and which may be ignored? What aspect of the experimental result is fortuitous; what portion due to the operation of the law for which we are searching. Answering such questions requires a guide, a cookbook, and this guide is normally provided by the scientific and extrascientific knowledge the investigator has already compiled, in school or in his research.

But such guides, although science cannot proceed without them, may also be blinders. For they are undoubtedly just such elements as those Pearson labels prejudice and preconception. They are as likely to direct the scientist's attention away from the fruitful observation as toward it. They represent habit and conservatism. Yet they cannot be dispensed with. For scientific observation is always a process of abstraction. One abstracts the length, the color, the texture from a natural object which always provides an infinity of alternate abstractions. Some choice is demanded, and the choice must ultimately rest upon personal prejudice.

I began this evening's lecture by indicating my conviction that elements of this sort are absolutely essential to the fruitfulness of creative research. And I hope that in concluding it I have provided a preliminary indication of the sources of this conviction. But if they exist, and if, as I suppose, they provide both direction and structure to every creative research effort, then the study of these preconceptions should itself be an important part of the study of scientific method. For their existence raises problems that the methodologist cannot ignore.

Is there some one set of preconceptions, some one criterion which is preeminently scientific? Should we train scientists in this one and this one alone? Is it strict adherence to his set of preconceptions which truly constitutes the "self-elimination" in judgment, which for Pearson characterizes the scientific man? Or can any set of preconceptions prove fruitful? Is the creative scientist actually the man who most strongly displays his individuality of judgment by proceeding from preconceptions different from those of the majority of his profession? And if so what are the sources of these new prejudices? How complete is their domination of research; by what can they be altered?

On Pearson's methodology these questions do not exist, or if they do exist the answers to them do not illuminate the nature of science. If this evening's lecture has cast doubt upon the utility of so pure an empiricism, if it has evoked in you any prejudice in favor of this alternate direction of research, then it has more than achieved its purpose.

Notes

1. In Kuhn's lecture script, "Mr. Lowell" appears crossed-out and bracketed by hand.

2. "Pearson 6." Here and in other editorial notes I reproduce the parenthetical references in Kuhn's scripts in quotation marks. In this case, he referred evidently to the second edition of Pearson's *Grammar of Science* (London: Adam and Charles Black, 1900), page 6, but omitted Pearson's emphasis in the second quoted sentence which extends from "The classification of facts" through "the function of science."

3. "Pearson 9." See Pearson's *Grammar of Science*. Kuhn's quotation omits Pearson's recommendation and hope that "such a man . . . will carry his scientific method into the field of social problems"—an area of concern that Kuhn briefly addresses later in this lecture and excludes from methodological consideration.

4. "*Novum Organum*, Aphorism 82." See *The Works of Francis Bacon*, vol. 4 (London: Longman and Co., et al., 1858), Aphorism 82, page 81. Ellipses inserted where Kuhn omitted words or phrases. Kuhn's quotation contains the errors "deducing axioms" (instead of Bacon's "educing axioms") and "the created man" (instead of Bacon's "created mass").

5. ". . . which govern them" has been rendered "which governs them." Elsewhere Kuhn speaks of both singular and plural laws governing some range of phenomena, so he could have meant either.

6. Kuhn's script at this point includes a handwritten insertion reading "Both contribute—but not the same."

7. For an account of Kuhn's self-described "acute distress" over this advertisement, see my *Politics of Paradigms* (Albany: SUNY Press, 2019), ch. 9.

8. In fact they fall at the same rate of acceleration and therefore with the same increasing speeds. Kuhn should have written (and perhaps intended to write) "...same speeds," such as those speeds recorded by the gaps marked by the experimental apparatus he describes.

9. Kuhn partially underlined by hand the phrase "performed by any man."

10. At the end of this sentence Kuhn inserted by hand and in capital letters, "Bad History." Because this is not by itself a sentence, I have excluded it from the main text here, but the words do appear below.

11. "Cooper 55." See Lane Cooper, *Aristotle, Galileo, and the Tower of Pisa* (Ithaca, NY: Cornell Univ. Press, 1935).

12. Kuhn provided no source for this quotation, but it can be found in Mersenne's *Harmonie Universelle* (Paris: 1936), 112.

13. "TNS. 171." The quotation appears on page 179 in the translation by Henry Crew and Alfoso De Salvio of Galileo's *Dialogues Concerning Two New Sciences* (New York: Macmillan, 1914).

14. For a more complimentary view of Galileo's experimental abilities and results, see for example, Stillman Drake, *Galileo At Work* (Chicago: University of Chicago Press, 1978).

15. Here Kuhn inserted by hand "Newton & Kepler."

16. "Randall 235." See John Herman Randall Jr., *The Making of the Modern Mind* (New York: Columbia University Press, 1976 [orig: Haughton Mifflin, 1926]) .

17. "Koyre III-23." The dialogues in question are Galileo's *Dialogues Concerning the Two Chief World Systems*, but Kuhn quotes and translates from Alexandre Koyré's *Études galiléennes* (Paris: Hermann, 1939), vol. III.

18. Kuhn originally typed "For what were these 'facts'?" Pearson, *The Grammar of Science*, 9.

19. At this point, Kuhn's typed script is attached to an inserted slip of paper on which is typed what appears to be a short, concluding paragraph that substitutes for the last twelve paragraphs of this lecture, beginning at this point. This shorter ending reads:

> Instead, if I may anticipate the material of my next lecture, we shall discover that the "facts" from which Galileo drew his conclusions were vague facts, qualitative facts, entirely lacking in numerical precision. And we shall see that from such qualitative facts any one of a number of technical conclusions might be drawn:—from the same facts Aristotle did draw quite different conclusions. And this will confront us with the necessity of determining both the differences and the sources of the differences in the attitudes or preconceptions with which Galileo and Aristotle approached these facts. I hope it will prove an illuminating study.

Lecture II

The Foundations of Dynamics

Toward the end of the last lecture, I suggested that an examination of the fable of Galileo presents us with the problem of determining the manner in which Galileo arrived at his novel laws, and I mentioned that the particular interest and difficulty of this problem lies in the fact that Galileo's novel laws were produced without any correspondingly novel observations or experiments. It is to this problem that I should like to address myself this evening. Why did Galileo see one set of laws implicit in a set of observations from which his predecessors had derived another?

This question in turn cannot be answered without an examination of the manner in which scientists before Galileo had rationalized these observations, for Galileo, genius or not, did not approach the problem of motion without education. And this education provided a set of opinions about motion which had satisfied a number of learned men but which Galileo rejected in favor of his own novel laws. And we will have to ask what these opinions were and why he found them inadequate. At least we will have to ask this question if my last lecture convinced you of the fallacy of the usual answer: that Galileo rejected Aristotelian physics because he discovered by experiment that it did not work.

The single most difficult thing to grasp about the problem of motion in ancient and medieval thought is that it was not the same problem to which we and Galileo apply the term. The problems which we label as problems of motion or dynamics were for Aristotle simply a special subgroup of the larger set of problems presented by the existence of change. For the perception of alteration raised for the Greeks and still raises for some modern philosophers a dialectic problem of the greatest difficulty—a problem rooted in the dilemma of analyzing flux in terms of logical categories which are themselves outside of time. In philosophical terminology this is the problem of retrieving "being" in a world primarily characterized by "becoming."

Before Aristotle's time, the Greek philosopher Heraclitus had pointed out that all perception is perception of a continuous alteration, and had reached the Bergsonian conclusion that since all is change there can be no permanent entities. The philosopher Parmenides had adopted the opposite alternative, and had argued that nothing which does not exist can come into being and that no existing entity can pass away. From this dialectic he had concluded that alteration observed in the world is but appearance, and that reality is but one and is changeless.

Neither of these solutions which avoided one horn of a logical dilemma only by grasping the other with acute discomfort was satisfactory to Aristotle, whose essential objective was the reconstruction of the commonsense world. But though he rejected these conclusions, he could not reject the problem, and the effort to find a satisfactory solution for it conditions a large part of his thought and that of his medieval successors. We can here ignore the dialectic portions of Aristotle's solution, but we must deal with the manner in which it affects his analysis of motion. And the first and most important of these effects is that Aristotle treats a motion as a change analogous to all other alterations of the physical world. The word motion, for him, embraces changes of color and size and form as well as those of position. In his own phrase: "The fulfillment of what exists potentially, insofar as it exists potentially, is motion, namely, of what is alterable qua alterable, alteration; of what can be increased and its opposite, of what can be decreased, increase and decrease; of what can come to be and pass away, coming to be and passing away; of what can be carried along, locomotion."[1]

Now if all changes are thus considered to be analogous behaviors, if change of position is like change of size, then one must search for laws which will apply to all these sorts of change together. And the primary aspects of nature to be abstracted in the analysis of any one alteration are accordingly just those which all alterations have in common. It is in these terms that Aristotelian analysis proceeds.

In the first place all motions or changes can be divided into two mutually exclusive categories: they may be *natural* alterations, changes which follow from the nature of the object moved and which do not perturb the normal processes of the physical world. Or they may be *violent* motions, produced by an external force which disrupts the natural course of events. The growth of the acorn into an oak is a natural motion; the destruction of the oak by the woodsman's axe is a violent motion. The upward course of fire when, freed from the imprisoning log, it leaps upwards to its place in the heavens is a natural motion; when the fire is constrained by a chimney or roof the motion is violent.

This classification may be applied equally successfully to the phenomena which we call dynamics or the movements of heavy bodies. Natural motions are the motions which occur automatically in the absence of a restraining force. The typical natural motion is that of the falling stone which, after its release from the restraining hand, rushes to regain the natural position of a heavy body near the center of the universe. Violent motions are those which require an external force, a shove or a push. They are typified by the movement of a stone raised from or dragged along the ground, or by the flight of a projectile hurled from the hand or a sling. And these, like other violent motions, are disruptive. They deprive the heavy body of its natural state which is the state of rest, close to or at the center of the universe. In the dynamics of Aristotle and the early scholastics these two sorts of motion are always separated. They cannot occur simultaneously in the same body, and they do not normally obey the same laws.

It is worth noticing that this distinction between natural and violent changes is one given directly in everyday experience, and that it is a distinction which we still apply to many of the subjects which Aristotle discussed under the single rubric *motion*. The terms violent and natural are of course out of style, but in daily life and in many sciences it has again and again proved profitable to distinguish the normal from the abnormal or pathological, and to adopt different modes of thought and different techniques of analysis in approaching these two aspects of experience.

But though the distinction is founded upon experience and has been pragmatically justified by the fruitfulness of its application in many fields of research, it turns out to be a block to progress in the study of the motion of terrestrial objects. And we are going to examine the manner in which it was removed. But first we must note another consequence of the Aristotelian analogy between motion and change. And this may again be approached in

Aristotle's own words: "Every change," he says, "is from something to something, as the word itself (*metabole*) indicates, something after something else, that is to say, something earlier and something later."[2]

To paraphrase this in a somewhat more leading form, every change is to be understood as a change from some initial state to some other final state. Any particular change is completely categorized by the difference in magnitude between these initial and final states, and when the speed of a change is discussed, it is taken to be the total amount of change divided by the total time taken to produce the change. Thus, to return to locomotion, if a body is moved from one end of this rostrum to the other, its speed is taken to be the total length of the rostrum divided by the total time it has taken to cross it; or if a body falls from the edge of this platform to the floor, the speed of its fall is again the height of the platform divided by the time which the body has taken in traversing the distance from the platform level to the floor. This, of course, is the figure which we should call the average speed.

The assumption that the important characteristics of any motion are the distance between its endpoints and the total time required to traverse this distance lead, when applied to everyday experience, to a number of qualitative laws of motion and indirectly to similar quantitative laws. For example, to move a block of wood from a position at rest at one side of this rostrum to a position of rest at the other, an external force is required; for this is a violent motion. If the force is greater, the time required for the transfer will be less. If the body to be moved is lighter, the force required will be smaller. And if the resistance to the motion, due either to friction or to the medium through which the motion is effected, is greater, then the time required to complete the motion will also be greater.

These are general qualitative laws about the forces required to move real bodies through particular intervals in definite times. They are not in the least arbitrary; on the contrary, they provide an accurate description of facts with which we are all familiar. It is easier to shove an empty trunk than a loaded one across the floor; in either case, the movement is easier when the floor is smoother. And this is the qualitative content of the Aristotelian laws for violent motion. They are accurate, and they are complete. That is, they are complete, if a motion is a change, if it is a transition between fixed endpoints.

But they are not quantitative laws, and since it is in their quantitative form that they are most often subjected to criticism, we shall now have to observe the techniques employed by Aristotle in strengthening them so that they specify the amount of time required for the motion of a particular body of a particular distance under the influence of a particular force. The quantitative laws are not of much importance in the totality of Aristotle's thought. He introduces them almost as asides, in forms which are not always either clear or consistent, and he never applies them. Nevertheless we cannot ignore them as he so nearly does, for they display an important type of scientific thought and they provided a point of departure for Aristotle's scholastic critics.

The essence of Aristotle's quantitative procedure lies in assuming the simplest sorts of quantitative relationships between the variables he has already isolated, that is, in assuming the simplest quantitative relationships which will supply the qualitative behavior already contained in the old laws. For example, since the time required to complete a particular motion decreases as the force increases, Aristotle supposes that this decrease in time is directly proportional to the increase in force. He says that if the force on a body to produce a particular motion is doubled, the time required for that motion will be cut in half. And similarly, he says that if the force remains constant but is applied to a body with only half the weight, then only half the time will be required for the motion. The resistance of the medium is treated in the same manner: if the medium is half as dense, then the motion of the same body under the same force will be twice as rapid.

These quantitative laws are of course invalid. If we were to test them with modern laboratory equipment, we should not get the predicted results. But one would get qualitatively correct results which is as much as anyone could get with Galileo's laws before the middle of the seventeenth century. And Galileo's laws predict results which we can today confirm. In any case no such quantitative tests were applied. The laws were criticized not for lack of agreement with the experiment, but because of logical difficulties which arose in the attempts to apply them to motion considered purely abstractly.

But what of the procedure by which the quantitative laws were arrived at—the assumption of the validity of the simplest quantitative relationships which would yield the correct quantitative behavior? By modern standards it is highly irresponsible. But it cannot be rejected out of hand as unscientific, and it is identical with a procedure Galileo used in arriving at correct laws of motion. But for Galileo motions meant something else, and we shall have to see how this change occurred.

So far I have said nothing about natural motion. All the laws discussed so far are laws for violent motions; they do not apply to the case of the falling stone. Indeed, about the problem of free fall, Aristotle himself had nothing explicit to say. Some of his followers seem to have assumed that if, in this natural motion, the acting force were equated with the weight of the fallen body, then the laws for violent motion would apply. And in this case it would follow that if a 10 pound weight and a 1 pound weight were dropped together, the heavier would reach the ground 10 times more quickly than the lighter. But Aristotle did not say this, and I am not sure that any philosopher or scientist ever believed it. This law seems to have been attributed to Aristotle primarily by his critics.

This was all that Aristotle had to say about the motion of terrestrial bodies. These laws covered for him what were in his day the important characteristics of this sort of alteration. But it was not all that he had to say about physics, and we shall not understand the way in which his laws were overthrown unless we know how closely they were associated with a set of views about the cosmos which were rejected together with his laws during the latter portions of the middle ages. For Aristotle's laws, like all scientific laws, did not stand, or fall, alone. They were an integral portion of a larger fabric of natural philosophy for which they provided support and by which they were supported.

This evening we can only touch upon two isolated features of this larger cosmology: the Aristotelian universe was a small universe, and it was a full universe. At its outermost edge it was bounded by a crystalline sphere in which the stars were set. Within this sphere and concentric to it were some 40 smaller spheres, set surface to surface like the layers of an onion. These spheres carried the planets, and their interlocking rotary motions accounted remarkably well for the major astronomical observations of the day. At the very center of this nest of crystalline spheres, at the mathematical center of the universe, was set the spherical earth, and the space between the surface of the earth and the innermost sphere which carried the moon was filled by the elements air and fire.

Both psychologically and scientifically it was a satisfying model. Until Galileo turned his telescope on the heavens at the beginning of the seventeenth century, the important features of this cosmos were confirmed by the bulk of observational evidence. And the model had tremendous psychological consequences for the study of motion. In the first place, every moving body moves through a space which is already full; if it moves in the region between the earth and the moon, it moves through air or through fire. And the medium through which it moves inevitably resists the motion, just as water causes a drag which slows the motion of a boat. So any student of motion must consider both of the forces active in producing it: the moving force of the pusher and the resisting force of the medium. One might discuss the ideal case of movement in a vacuum—Galileo of course does just that—but since in the Aristotelian universe there cannot be such a thing as a vacuum, the discussion of a motion which was not impeded by a material medium seemed without practical or theoretical significance. The abstraction to motion through the void was an unreasonable abstraction. It is as though a modern physicist were asked to study the fate of a cannon ball fired with the velocity greater than that of light.

In the second place, the very small size of the sublunar region in which terrestrial qualities could move gave the problem of motion an aspect quite different from that of motion in an infinite Newtonian universe. The notion of an infinite continuing motion, except for circular motion of the heavenly bodies, was, in so confined a space, another unreasonable and impractical abstraction. Every motion must be a motion from one place to another place. And the particular places involved were themselves important.[3] For this bounded sublunar region in which the terrestrial motions occurred had the psychological aspect of a room or a hall with whose furnishings the philosopher was familiar. The physical bodies which filled it were its furniture, and each had its own unique natural place of rest. A natural motion was a motion toward this place, as that of a stone toward its place in the center of the universe. And it followed from this understanding of space that important characteristics of a natural motion were determined by the natural position of the body moved, by the particular endpoint toward which alone this particular natural motion tended. So that once again the endpoint, the final state of rest, was an important determinant of the motion of terrestrial bodies.

The Aristotelian doctrines relating to motion represent the highest development of this facet of antique thought. Further progress, which had in fact commenced, was cut off by the series of wars which brought Hellenic civilization to its close. But coincident with the reintroduction of Aristotelian writings into Europe in the thirteenth century developed a tradition of Aristotelian criticism known as scholasticism, a critical tradition which immediately introduced important modifications of the Aristotelian theory of motion. It was in this modified Aristotelian tradition that Galileo was trained, and if to us he is usually portrayed as the first modern student of motion, he may with equal justice be viewed as the last of Aristotle's scholastic critics. For Galileo's work is not a beginning but a turning point. Implicitly it was within the scholastic tradition, but its outcome so completely modified the tradition that his successors were able to break with it altogether.

The scholastic contributions leading to Galileo were the products of two independent critical traditions, one logical and the other physical. Those who criticized Aristotle on physical grounds attacked not the foundations of the theory but incidental remarks which Aristotle had made about two special problems, the problem of the projectile and the problem of the falling body. Aristotle had said that every violent motion requires a pusher. And this led to the greatest difficulty when he considered the problem of the stone hurled with the hand or from a sling. Obviously such a projectile continues to move after it has left the hand, so one may legitimately ask, what is pushing the stone?

The difficulty was recognized by Aristotle, who produced an answer to it which was universally unsatisfactory to his critics. After the stone has left the hand, he said, the air through which it moves continues to push it along, and he supplied two possible mechanisms for this continued motion. Neither of them was intrinsically impossible. Either could have been subjected to a consistent and coherent logical development. But they appeared unreasonable, both to Aristotle's contemporaries and to his scholastic followers. For the air is at best a clumsy intermediary, and it seems superfluous. If the hand can store up a continued push in the air, why can't it store the same push in the projectile itself. It was in fact this suggestion which, revived in the thirteenth century by Albertus Magnus, was erected into a new science of motion by the Parisian Nominalists in the century which followed. This new science received the name of the impetus theory, for it is held that during the initial motion of the hand an impetus, an internal moving force, is stored in the body, and after the projectile leaves the hand, this impetus continues to push it along. In this continued pushing the impetus is itself exhausted until finally the body falls to the ground.

The new theory explained a great many things. A body can be thrown farther from a sling than from the hand because it is in contact with the sling for longer, so that more impetus is stored up in it. Or again, a heavy body, a stone, can be thrown a great deal farther than a lump of parchment, for the heavy body can receive more impetus from the hand than can the lighter one.

But these limited insights into particular problems of motion were not the most important consequences of the impetus theory. Far more significant was the reorientation which it produced toward the nature of the problem presented by terrestrial motions. The impetus theory concerned itself with the fact of continued motion, and its concentration upon the manner in which impetus is stored in the body and is exhausted during the course of the motion directed attention to just that aspect of motion which the Aristotelian theory had ignored. It directed attention to what goes on during the course of that motion.

Motion for Aristotle was simply change, and as change, it was represented by its initial state of rest and its final state of rest. Position was for Aristotle a quality of the object, and a motion, a change in position, was a change of quality comparable to a change of color. But with the development of impetus theory, the impetus itself became a quality stored in the body. It was compared with sound, stored in a struck bell, or with heat stored in a warmed metal. And motion thus ceased to be a simple change of quality, and became very nearly a quality, a state of the object. To be in motion, as to be yellow, was to be possessed of a quality. To study motion was to study a continuous process, not merely endpoints.

This transition in the attitude toward motion was not completed in the medieval period. It was first fully formulated in the seventeenth century by Descartes, who was also the first to announce its physical corollary, the conservation of linear velocity, or the law of inertia. But the impetus theory, proceeding from the limited criticism of the Aristotelian explanation of the motion of the projectile, represents a major step toward the modern view, for by forcing attention to a previously irrelevant aspect of the motion, it separated the problems of dynamics from those of qualitative change, and it forced medieval philosophers to introduce a new set of categories applicable to locomotion alone.

The effects of this new conception of a motion as a state rather than as a change are clearly seen in the position taken by the scholastics with respect to the problem of the falling body. On this problem too, the position attributed to Aristotle had been subject to criticism almost from the time it was enunciated. The scholastics now suggested that this sort of motion could be explained by the impetus theory: a body commences to fall because of its weight—weight which was now regarded as producing an internal force like impetus. But since the weight acts not only at the beginning of the motion like the hand, but throughout the motion, the body in falling must continually acquire more and more impetus. Thus its speed increases as it falls.

Some fourteenth-century thinkers, pursuing this line of thought, noted that since the impetus must increase uniformly during the motion, it was probably that the velocity increased in the same manner, and they thus came very close to the law of uniform acceleration with which Galileo is credited. Some of their writing can be interpreted as a literal enunciation of this law, but scholastic thought on the problem is by no means clear.

For us this priority question is of no importance. What we must not miss, however, is[4] that the impetus theory had directed attention for the first time to the existence of such a thing as acceleration. By studying motion as a state of existence, the new theory had brought about the recognition of changes of this state. Aristotle, who was a shrewd observer, can scarcely have been unaware that a body moves more rapidly as it falls, but nowhere does this observation appear in his *Physics*; for it was not a relevant characteristic of motion if motion was considered to be change between positions of rest.[5]

The impetus theory, then, led to a concentration upon new aspects of an old problem. But this does not exhaust its importance. Equally significant is the way in which it blurred distinctions which had previously been important. Note in particular the manner in which the difference between natural and violent motions has disappeared. Both of these are now accounted for by the common quality of moved objects—impetus. A stone flung vertically gradually loses the initial impetus upward, impressed upon it by the hand, and simultaneously gains impetus downward due to the continuing effect of its weight. It thus loses speed until it hangs stationary at the top of its trajectory after which the impetus downward dominates the motion, and the body falls with a constantly increasing velocity.

Here the violent upward motion and the natural motion downward appear completely symmetrical. Gradually it was recognized that the impetus gained on the way down will be just the same as the impetus lost on the way up. The body hits the ground with the same velocity with which it left the hand. In the sixteenth century such considerations were applied to more complicated trajectories, particularly those of canon balls, and before the time of Galileo some military engineers had realized that not only must natural and violent motions be considered as one problem, but that the two motions could be combined to produce a continuously curved trajectory. For Aristotle, you'll remember, natural and violent motions were incompatible; they could not coexist. Nor, for the same reasons, could curved and linear motions be compared. They were all different sorts of changes, as different from one another as from changes of color.

Finally the impetus theory did a great deal to suppress the importance of what in Aristotelian physics had been absolute place. All of Aristotelian space was a position for something, and the character of at least a natural motion was largely determined by the place toward which it tended. But for a scholastic defender of the impetus theory, the characteristics of the motion were primarily determined by the amount of impetus put in at the beginning. A motion characterized by the same initial impetus and by the same rate of gain or loss of impetus would proceed in the same way regardless of the position in the universe in which it was executed. Thus, in the impetus theory, position relative to the point at which the impetus was delivered, rather than position in absolute space, became the important characteristic of a motion. And in the application to the freely falling stone this transition made the important variable this distance through which the stone had fallen, rather than the distance between the stone and the earth. This, of course, is just the transition required to give the law of free fall a mathematically simple form.

Thus the impetus theory produced a small but important step toward what we should now call the relativity of inertial motion, and this trend was very much reinforced by the astronomical speculations which occurred during this and later periods. From the fourteenth century on, a number of important scholastic thinkers suggested, normally on a speculative or mystical basis, that the earth was like the other planets and that it moved, and these developments provided a conceptual basis for the technically more complete astronomical revolution proposed by Copernicus in the sixteenth century.

But if the earth moves, then the motion of a falling body is toward the earth rather than toward the center of the universe. The motion may no longer be explained by the body's preference for some particular point in an absolute space but must depend on the body's position with respect to the earth or with respect to the source of its motion. Similarly, if, as the Copernican theory insists, the universe is much vaster than Aristotle thought, or if it is infinite, as many followers of Copernicus suggested, then much of the psychological force of absolute position is lost. For the notion that every point in space is specifiably physically distinct from every other point is psychologically tied to the belief that the total amount of space available is rather small, and that the totality of space can be grasped rather like the totality of space in a room. Again, in so large a universe, it will make sense to talk about an infinite motion in a straight line and thus to arrive at the abstraction generally stated as the law of inertia. In an infinite universe which, by Galileo's time, was a subject of common speculation, an infinite motion in a straight line is a reasonable abstraction, while to Aristotle it had appeared inherently contradictory.[6]

Thus the impetus theory, which by the end of the sixteenth century had replaced the older Aristotelian theory in most of the major educational centers of Europe, provided a new conceptual framework within which terrestrial motions were considered. A motion had become new sort of conceptual entity; its important characteristics were no longer the same as they had been for Aristotelians. And the transition had occurred against a background of new cosmological speculation and of radical astronomical theory which reinforced the new viewpoint. By the end of the sixteenth century the very meaning of the word had changed. In the fourteenth century as in antiquity the word *motus* meant change or alteration, but when Galileo at Pisa writes a treatise called *De Motu* he addresses himself to a field which only began to exist in the fourteenth century—the field of terrestrial dynamics. Many of the attacks which he directs at Aristotle from this new vantage point miss entirely the point of what Aristotle had to say. He is talking about a different problem and using the same words in different senses.

And by Galileo's time two[7] other important changes had occurred. The first of these concerns the role played by the medium in terrestrial motions, and we shall have more to say of the sources of this change in the next lecture. Here we need only note that by the sixteenth century a number of philosophers and engineers had grown very dubious about the logical impossibility of a vacuum, and that some, including Galileo, thought that a vacuum might be produced in nature by a finite force. Thus motion through a vacuum became, for some thinkers, a possible abstraction and a significant ideal case.

The second development is the one I referred to earlier as the logical criticism of Aristotle. It derived not from the consideration of particular motions but from the apparently separate problem of the logical categories to be applied to the analysis of any change, and it was carried on independently of the physical criticism of the impetus school. It originated at Oxford after the start of the impetus school on the Continent, and although it was occasionally advanced by men who also made important contributions to the impetus theory, particularly by one Nicholas Oresme, the relevance of the logical problems of the Oxford school to the physical problems of the Parisian school was scarcely noted before the time of Galileo.

As we have already noted, there are sound psychological reasons for supposing that the essential character of any alteration is given by a knowledge of these two endpoints of the change; we shall see later that many children handle problems of alteration in exactly this manner. But to convert this sound insight into a logical doctrine and to say, as Aristotle does in many parts of his writing, that the change is identical with the two endpoints or with the final endpoint of the alteration, assimilates the essential characteristics of a change to those of a state of rest. And this creates internal contradictions in any attempts to analyze a particular motion. For as the paradoxes of Zeno had indicated in antiquity, the consideration of a change as a succession of states or rest demands the analysis of an infinity of such states, and the Aristotelian categories were not well suited to treat such a problem.

The new logical categories developed were both obscure and abstract in the highest degree. Even the preceding statement of them contains simplifications which depart from the spirit of the medieval problems. But if we cannot follow its detail we can note that the outcome of continued fourteenth-century attention to the logical analysis of change was the creation of a new set of logical and mathematical tools which could be applied to problems of change in general and of motion in particular. For example, fourteenth-century scholars solved a number of problems of this sort: if a body starts from rest with a velocity of one mile per hour and maintains this velocity for an hour, and if it then travels for the following half hour at twice this velocity, and for the following one-eighth of an hour with eight times this velocity and so on, how far will it have gotten at the end of a two hour period?

And among the problems treated in this manner was the problem of a body which starts from rest and travels for a limited time with a velocity or speed which increases uniformly and continuously during this time. This is the problem which we should now call the problem of uniform acceleration, and the fourteenth-century logicians arrived at the modern solution: a body starting from rest and proceeding with uniform acceleration, covers, in a given amount of time, the same distance as that covered by a body travelling with a uniform velocity equal to one half of the final velocity of the accelerated body.

Here I am guilty of an overly modern statement of the outcome of scholastic research. Their problems are problems of change in general, not simply problems of motion, and the notion which we should now describe as that of instantaneous velocity is by no means explicit in their analysis. That notion is not even clear in the work of Galileo, who three centuries later borrowed quite literally the fourteenth-century analysis made by the scholastic Oresme. And a glance at the confusions, inconsistencies, and misunderstandings which so often characterize these medieval logical analyses leaves no doubt of why this brilliant but incomplete work should have waited almost three hundred years for its fruitful application.

But whatever its inadequacies and confusions, the work did make abundantly clear certain logical aspects of the problem of motion without which the modern analysis, or even Galileo's analysis, of motion would have been impossible. It led, for example, to the recognition that a number of motions which cover the same total distance in the same total time may accomplish this single transition by an infinity of different processes. Thus it demonstrated that analysis of motion requires a knowledge of the intensity of the motion at each instant of the transition, that is, demands an understanding of what we should now call instantaneous velocity. And the new logical school created graphical and algebraic tools which permitted this newly isolated variable to be studied. It was in the application of this new logical understanding and these new logical tools to the qualitative physics of the impetus school that Galileo made one of his principal contributions—a contribution whose ultimate outcome was of course to destroy totally the impetus school itself.

Which brings us to the work of Galileo. An almost trivial example of the way Galileo applied the new logical and physical insights is presented by his enunciation of the law that the distance through which a body falls from rest is proportional to the square of the time during which it falls. This law had been proclaimed earlier, at least by the Spaniard De Soto in the sixteenth century, but Galileo's independent development of it was more complete and more widely circulated, so that it was to his work that all his successors referred. And the secret of the new solution was his new statement of the problem. With the benefit of his training in the impetus theory Galileo grasped the problem of free fall as that of describing the motion of a body in a continuous state of motion travelling away from an initial point of rest with a velocity or intensity of motion that increased uniformly throughout.

With the problem stated in this new manner, a new solution presented itself almost immediately. Galileo described the process as follows:

> When therefore I observe a stone initially at rest falling from an elevated position continually acquiring new increments of speed, why should I not believe that such increases take place in a manner which is exceedingly simple and rather obvious to everybody? If now we examine the matter carefully, we find no addition or increment of velocity more simple than that which repeats itself always in the same manner. [. . .] Thus we may picture to our mind a motion as uniformly and continuously accelerated when, during any equal intervals of time whatever, equal increments of speed are given to it.[8]

But this uniformly accelerated motion was simply a special case of the more general sorts of change considered by Oresme. And it is from Oresme that Galileo borrowed both the demonstration and the result that a uniformly accelerated motion produces a displacement that is proportional to the square of the time of that motion. At a later date, perhaps in order to satisfy his critics, he devised the inclined-plane experiment to confirm this law, but the inclined plane could do no more than to show that the motion of a rolling body was more or less of the sort which Galileo had described. It showed that the distance was not proportional to the time as Aristotle would have supposed, but that in fact it increased rather more rapidly. The law itself, with all its precision, Galileo derived by supposing that the actual motion must obey the mathematically simplest law which would provide the observed qualitative characteristics of free fall. But this was exactly the method which Aristotle had employed in arriving at a different sort of law, and the phenomenon had not altered in the interim. What had changed was the scientific view of the phenomenon: motion had ceased to be a change between fixed endpoints and had become a quality of the moved body, a quality whose intensity was observed to increase throughout the motion.

A similar dependence on the scholastic reorientation toward motion is observed in Galileo's derivation of the law that the rate at which bodies fall is independent of their weight. Here again he employs experience only to tell him that Aristotle was wrong, a fact which had been noted at least occasionally in the European literature since the sixth century. And again he reaches the correct law by a mental rather than an experimental argument.

It is an argument of the greatest ingenuity. If, says Galileo, we consider the fall of two identical bricks released simultaneously from a high tower, we know that since the two bricks are the same they must fall side by side. But since they fall exactly together, it will not affect their motion at all if we tie them together to make a single brick. This new brick will be just twice as heavy as either of the original two taken alone, and it will fall at the same rate as either of the others. A similar argument will hold for three bricks or four bricks or any number of bricks, so that finally Galileo is led to the conclusion that any two bodies, at least any two bodies of the same material, will fall at the same rate regardless of their weight.

The extension of this derivation, which applies only to bodies of the same density, to the more general case, in which bodies of different densities are considered, is an intricate one made even more difficult by the errors Galileo committed in tracing it. But for present purposes we need only examine the first step already sketched, for although it sounds completely convincing to us, it contains a hidden premise which made it totally inadmissible to anyone regarding motion from the Aristotelian viewpoint. For the statement that tying the two bricks together cannot affect the motion is logically valid only to the extent that the motion is governed by a force independent of the size and shape of the body on which it acts. To the extent that the medium exerts an appreciable effect on the motion, joining the two bricks may alter its character entirely. The air will no longer rush through them and will therefore exert a larger resistance. If the air serves the additional function of a pusher as it does for the Aristotelian projectile, the two bricks combined will present a larger area to push. These two effects might cancel each other; Galileo's statement might be valid. But whether or not the result is correct, the argument is not, except for a person who already believes that the primary forces governing the motion reside within the body, and this point of view is of course foreign to Aristotle.

It is therefore no surprise to discover that arguments almost identical with Galileo's had been applied to the problem and rejected before his time. The critics examining them had declared that they were valid only to the extent that the motion of the body was governed by the body's weight, but that since for any actual motion, many other factors, particularly the medium, were important, no such conclusion would hold for real bodies in the real world. Again, Galileo's argument is dependent upon a new point of view about the essential characteristics of an old phenomenon.

Finally, let us turn our attention to the observations and discussions in which Galileo's greatest genius is displayed, the discussions of the pendulum. Here we are confronted with a totally new insight, one which is not, so far as I can determine, even hinted at in any of the previous scholastic discussions.

Galileo first discovered the important property of the pendulum, when, as a young medical student, he spent some time during a lengthy church service in watching the motion of some lamps suspended from the roof. By timing each swing with his pulse, he noted that even when the swings were small, they appeared to take the same time as the larger swings which had preceded them, and that when the wind again set the lamps in more violent motion, the oscillations were still performed at the same rate. He then, so the story goes, applied the pendulum to the determination of the pulse-rates of his patients, and presumably in the course of this work discovered that the period of oscillation is also independent of the material of which the bob is built.[9] He discovered that the time taken to complete a swing depends only on the length of the suspending cord.

On this occasion he let the argument from simplicity lead him astray. He reported that the time required for an oscillation was independent of the amplitude of the swing not only for small amplitudes but for very large ones as well, and this is not even approximately true. But even the error is of extreme significance. If Galileo had been a more careful experimenter he would not have committed it; but if he had been more careful and less intuitive he would probably not have thought significant the original crude regularity observed with the aid of his pulse in the windy church. And if he had observed the true complexity of the pendulum's motion, he would have been less apt to unify conceptually the motion of the pendulum and the motion on the inclined plane. For his mental unification of these two cases was dependent upon his seeing in them identical sorts of simplicity. And it is from this unification that Galileo's most fruitful contribution to dynamics proceeds.

The error was actually useful, for the importance of the observation is not that it provides a new quantitative law about the pendulum, but that it illuminated for Galileo the regularities of all terrestrial motions. It provided not a new law so much as a new sort of insight. And this, I think, helps us to understand why the observation was first noted and recorded at the end of the sixteenth century. If we ask, why it was Galileo who first made the observation, we can only receive the impenetrable answer, because Galileo was a genius. But if we revise the question and ask why the observation had to wait until Galileo's time, why it was not made by Aristotle, Roger Bacon, or Leonardo da Vinci, then I think we can supply a more illuminating answer. All of these men had seen suspended lamps swaying in the breeze; all of them had proclaimed the crucial importance of careful observation; but for them the important characteristics of the motion of the swaying lamp were not the same as they had become for Galileo.

For an Aristotelian the motion of a pendulum illustrates the manner in which a heavy body displaced from its natural position regains that position, closest to the center of the earth. It is a motion from a state of rest to a state of rest. It is just one of a large number of such motions, and in itself it is trivial. It supplies nothing new. For Galileo the oscillations of the pendulum are themselves a state of being. And the state is characterized by the continual repetition of the same oscillation. And this is a new phenomenon. The Aristotelian's observation is more nearly correct—the pendulum does in fact come to rest; it never repeats quite the same path—but Galileo's idealization is more revealing.

I therefore think it appropriate to conclude that these particular idealized abstractions, which alone are the abstractions which make the pendulum significant and worth recording, were more likely to be made by a genius at the close of the sixteenth century than by a genius in antiquity, for the motion meant something new and different to the later figure. For by the end of the sixteenth century the motion of the pendulum had acquired a significance which it could not have possessed until motion was considered a state, and until curved and linear motions were considered to be part and parcel of the same underlying phenomenon.

This difference in the meaning or significance of the same observation to our two geniuses, one from antiquity and one from the Renaissance, is,

I think, borne out by the manner in which Galileo puts his observations to work. For in spite of remarkable exceptions which we shall have occasion to examine in future lectures, it is usually true that a new observation, which is recorded and becomes a fact with which future scientists must deal, is observed and recorded by a scientist whose interests and training equip him to appreciate its significance, and Galileo certainly appreciates and exploits the significance of the pendulum.

From the remark that the pendulum regains the same height on each swing, Galileo proceeds to the observation that when a peg is placed vertically below the point of suspension of the bob, the pendulum still reaches the same vertical height on both sides of its swing, even though the length of the suspending cord is different on the two sides. Since for Galileo, as for other members of the later impetus school, the curved path of the pendulum is produced by forces of the same sort as those which produce the linear motion of a body rolling down an incline, Galileo suggests that this characteristic of the pendulum law be extended to the inclined plane. He suggests that a ball rolling down one incline and immediately going up another will, like the bob of the pendulum, reach the same vertical height on both planes, and by an elaboration of this argument he concludes that the velocity of a body rolling down an incline is dependent only on the vertical height of the plane, not on its angle of inclination or on its length. And this leads him finally to the conclusion that if the plane is perfectly flat and is not inclined a ball that is started on it with a certain initial velocity will continue rolling in the same direction with the same velocity forever. With this remark Galileo comes within a hair's breadth of the principle of inertia, the principle which, as Newton's first law of motion, has totally transformed physics.

It is by thus interrelating the properties of the pendulum with those of motion down an inclined plane and by applying the laws thus arrived at to limiting cases in which the plane is horizontal and the velocity is conserved or in which the plane is vertical and the body falls freely that Galileo made his most important and most original contribution to the study of motion. And it was on the firm foundation provided by this truly brilliant piece of research that Newton and his contemporaries were able to build a totally new science of motion from which all Aristotelian elements were finally eliminated.

With this approach to modernity, we may appropriately close this evening's lecture.

Notes

1. "201 A10, Physics." Here and below Kuhn quotes from the translation by R.P. Hardie and R.K. Gaye (Oxford: Clarendon Press, 1930), book III.

2. "224 B1 Physics." Book V. There are minor discrepancies between Kuhn's text and Hardie and Gaye's translation.

3. Kuhn bracketed the preceding three sentences and wrote "omit."

4. Kuhn bracketed the preceding paragraph and this paragraph up to this point, and inserted by hand "And here we note." The insertion may be substituted for the bracketed text, but Kuhn did not additionally mark the text for possible omission.

5. Kuhn wrote "1/2" in the margin at this point, possibly to record that reading his script to this point required half of his allotted time. Below, he writes "3/4."

6. The previous two paragraphs are bracketed, with an annotation "omit." There are also additional brackets and marks within these paragraphs, as well, the significance of which are unclear.

7. Kuhn crossed out "two" and inserted "an" before "other." This was possibly another time-saving measure allowing him to omit the discussion in this paragraph and move directly to what he had called "the second development" discussed in the following paragraph.

8. "TNS 154." Ellipses added. The quotation appears on page 161 in the translation by Henry Crew and Alfoso De Salvio of Galileo's *Dialogues Concerning Two New Sciences* (New York: Macmillan, 1914). At this point Kuhn writes "3/4" in his margin.

9. According to Stillman Drake, it was not Galileo but his acquaintance Santorre Santorio, a Venetian physician, who applied the isochrony of pendulums to the measurement of pulse rates of patients. Galileo's theories and observations of pendulums likely inspired Santorio's invention of a device to measure pulse rates. See *Galileo at Work*, 466.

Lecture III

The Prevalence of Atoms

In our last lecture we studied the transition in the attitude of Western scientists and philosophers toward a single problem, that presented by the study of the motion of terrestrial bodies. More precisely, we examined two different points of view toward the problem of motion and we saw how inadequacies in the first of these had led to a complete reformulation of the problem of motion and to a new set of laws. Tonight I should like to attack our subject differently. Instead of a single problem, we shall examine the application of a single approach to a variety of problems. We shall see, that is, the manner in which one metaphysical notion about the structure of the world has provided new insights to science and scientists.

This notion, commonly called Atomism, is the belief that the world is made up of an infinite number of microscopic particles that are in constant motion in an infinite void. This idea about the structure of the world is one we owe, *at least historically*, to the Greek philosopher Leucippus and his student Democritus who flourished in the Greek colony of Abderra in the fifth century, B.C.

I say "at least historically" for I should like this evening to find a middle ground between two extreme views which have been held regarding the relation of modern scientific atomism to the atomism of the ancient Greeks. The first is the truly absurd view that the Greeks, by sheer power of mentation, anticipated many or most of the conclusions reached by the combined efforts of nineteenth- and twentieth-century scientists. The second view, equally absurd, states that the resemblance between Greek and modern atomism is purely fortuitous; that it's simply a rather disagreeable accident that a lucky guess about the nature of the world by some obscure Greek philosophers should seem so similar to the totally different theory that emerged from careful experimentation and mathematical research during the past 150 years.

I should like to suggest as an alternative approach that this question is misphrased. For since the beginning of the seventeenth century, science has believed in and made use of a number of different atomisms. For there is no *one* scientific atomism. There are several versions and different fields of science have occasionally employed incompatible versions of atomism at the same time. But these various atomisms have evolved as they were employed in new scientific contexts. And through this evolution the basic notion of an atom has itself been changed.

So atomism has not been unaltered in its various contacts with science. But the same is true of science itself. This continued historical association of science and atomism has been fruitful. And if we trace the matter back, the first philosophical atomism to which science owes an important debt is the atomism of the ancient Greeks.

Greek atomism is in large part a product of the same set of dialectic problems that we examined in the beginning of the last lecture. Parmenides, you will remember, believed that everything which exists is eternal and changeless. From this it follows that the vacuum, whose very name implies the absence of all being, cannot possibly exist. For how can nonbeing exist? So there must be only one thing that exists—the real universe that is completely full, self-contained, and eternally changeless. This theory is the first philosophical monism that is of historical importance.

Aristotle responded to Parmenides's monism, you recall, by agreeing that there is no such thing as a void and that the universe is full. But, Aristotle recognized, this fullness of the universe does not imply an absence of change. Change and motion are still possible in a world without empty spaces as is illustrated by the motion of fish which move through water despite there being no spaces void of water; or the continued motion of a projectile which continues its motion through air because the displaced air itself, according to Aristotle's clumsy theory, pushes the projectile along as it rushes around to fill the void that threatens to appear in the projectile's wake. Aristotle's view dominated thought in Europe well into the sixteenth century—and this is not surprising given that this view of motion is built into Aristotle's system of knowledge and gives us a commonsense view of the world.

But Aristotle's was not the only way of understanding and preserving the reality of change from Parmenides's onslaught. Another way to understand change was available even before Aristotle and was contained in Leucippus's denial of Parmenides's reasoning that the very notion of the void was self-contradictory. On the contrary, Leucippus believed, the void exists, and so do little bits of microscopic matter that move about in it. It is the motion of these corpuscles that gives rise to the entire flux of experience.

What these corpuscles are brings us closer to understanding Greek atomism. For dialectic reasons similar to those which led Parmenides to conclude that the world was unitary, indivisible, and changeless, Leucippus and other early atomists held that these individual particles were similarly eternal and indivisible. Thus they were called *atoms*, where atom means undivided. In this way, on the basis of Parmenides's monism they made what can be called a *monadology*.

This atomistic view of the nature of reality had a number of logical and psychological consequences that were of great importance for later scientific thought. For one, there were now holes in nature because a genuine vacuum can exist. At the same time, it became clear that the universe must be infinite and extend forever in all directions. This conclusion followed logically from the first, for if there existed a boundary to the universe it would necessarily consist itself in corpuscles moving in the void. If we think there might be some area outside that boundary where nothing exists, then that area is simply a vacuum which is now granted to exist as a part of the atomistic universe. This is of course very different from the Aristotelian universe that I described in the previous lecture and which was understood to be finite and to consist in a set of concentric crystalline spheres. The earth was at the center of this system and the outermost boundary was thought to be a rotating sphere in which the stars were set.

More important for our present purposes, however, is the consequence of atomism that all the changes, all of the variable qualities that we observe in nature are produced by the changes in positions and relative motions of these fundamental particles. As the later atomist Epicurus said,

> We must suppose that the atoms do not possess any of the qualities belonging to perceptible things, except shape, weight, and size, . . . For every quality changes; but the atoms do not change at all, since there must needs be something which remains solid and indissoluble at the dissolution of compounds, (something) which can cause change; . . . changes affected by the shifting in position of some particle, and by the addition or departure of some other.[1]

This is a very interesting statement, for the ideas which it expresses have been a continuing source of two fundamental ideas about the universe which have continued to be a source of both inspiration and trouble for later sciences. This is partly because Epicurus's atoms are sensuously neutral. They possess only size and shape; that is, extension. As a result, to understand the variety and sensory luxuriance of the world exhibited to our senses we must not study our sense impressions, for they are scarcely trustworthy. We must instead search for the arrangements and motions of the corpuscles that underlie these appearances. This leads us to one fundamental idea, the tremendously important distinction between primary qualities which belong to the corpuscles themselves and the secondary qualities that we experience through our senses.

An even more important fundamental idea is that of the world as a machine operating behind our descriptions of the world whose parts are made of the same sort of stuff. Once the atoms are given their initial motions, they go on moving by themselves through the void and colliding with other atoms according to their own laws, not unlike the parts of a clock that, once wound up, continue to move of their own accord. Thus the universe itself could be seen as a giant machine, a piece of clockwork.

This idea is frequently thought to be a consequence of the Newtonian philosophy which dominated eighteenth-century thought. Newton himself, then, is often regarded as having single-handedly given the world this mechanical image of the universe. In fact, Newton did not entirely believe in this image himself. But as far as the popularity and influence of this notion is concerned, this claim about Newton is not misleading. This idea of the universe as a clock-like machine was nothing less than a dogma which influenced not only science and philosophy, but also politics and art.

For many working scientists, however, this notion of the universe as a machine is three quarters of a century older than the publication of Newton's *Principia*. And in this older idea we see an unequivocally Greek source. For at the beginning of the century Francis Bacon had written a defense of the works of Democritus and other Greek atomists. One of the main tasks of the new science, Bacon insisted throughout his life, was to study the primary motions which underlie the secondary qualities present to our senses. This is illustrated in Bacon's research, chiefly into heat, and which offered examples drawn from the Roman atomist Lucretius.

Soon after Bacon, René Descartes, the French philosopher whose work greatly influenced the continental science of the seventeenth century, provided a complete model of the world as a machine governed by invariable, God-given laws. The building blocks of Descartes's universe were little corpuscles moving through space and acting on each other only by impact. Descartes's views were so similar to ancient atomism that he was in fact accused of simply cribbing his philosophy from the Greek atomists. I could name others who were so influenced but will point out here only that the very name given to the major scientific tradition of the seventeenth century—"the New Philosophy" or "the Mechanical Philosophy"—seems in large part to have denoted an attempt, with the aid of experimentation, to reduce all phenomena to the motions of elementary corpuscles. Robert Boyle, in many ways the leader of the movement for the New Philosophy, was admittedly indebted to the Greek atomists. In his writings Boyle describes the universe as "a self-moving engine" and "a great piece of clockwork." Boyle's major work is an attempt to apply these notions to chemistry.

So far, we have dealt with atomism as a purely speculative cosmology that was drawn by the free, creative power of the human mind from the consideration of logical or pseudological problems. You may wish therefore to deny the name of science to this enterprise, and you may wonder why I am going into it in lectures devoted to the study of problems in scientific method. But I think you'd only be partly correct to deny the name of science to this material. I will discuss my reasons for this in the fifth lecture and should not like to argue this point now. Here I wish only to point out that whether or not it is science, it has had important effects on areas of research that are indubitably scientific, as we shall now examine.

The first of these examples is drawn from material in the last lecture, The Foundations of Dynamics. I should like to point out an important effect upon the study of motion of the recovery of a limited portion of the atomistic viewpoint. Aristotle, you will recall, conceived of a universe that is full, so that motion is always through a plenum. As a result of this, there were always two forces involved in any motion: a pusher and a medium. This view was broadly held until the sixteenth century, at which time scientists, while not fully embracing the atomism of Democritus, increasingly came to reject Aristotle's plenum. Instead, many scientists came to believe on philosophical grounds that there exist tiny vacua in things. These vacua accounted for phenomena like condensation and rarefaction. Very light substances like air, on this view, are highly rarefied and consist almost entirely of vacuum.

This means that motion through air is more like the motion of Democritus's atoms through the void, which offers no resistance to motion, than Aristotle's motion through a plenum which requires the constant efforts of a pusher. On this view, it became profitable for scientists to think about motions in which there is no resistance to the moving force. Motions caused by one force, in other words, became a worthwhile abstraction. This is the background for the acceptance of Galileo's new argument about the two bricks being tied together and falling at the same rate as the two bricks not so connected. When the medium was thought to play a different and important role, this argument was more readily rejected than it would be when the role of the medium came to seem less important.[2]

But the effect of this new attitude toward the void is not only felt in dynamics. The idea that vacua exist and that their size can be altered by finite forces suggested a new problem that was presented by an old phenomenon: the problem of pumps. It was known for more than a century before Galileo's time that pumps in mines would not raise water to great heights. Generally, they could not raise water more than 30 feet. This was not surprising given the materials and techniques available at the time. Pump shafts were made of wood, for example, and boring techniques were crude. On paper, however, engineers continued happily to design pumps that will raise water hundreds of feet. The idea never arose that there is an inherent limitation here. On the contrary, given the belief that there can be no vacua in nature, it followed that if a plunger is drawn up water must follow it (in the absence of leaks, and leaks certainly existed).

Faced with this situation today we'd eliminate most of the leaks and discover that these improvements scarcely increase pump effectiveness. This path was not available to Galileo because leak-proof tubes thirty-five feet in length could not be fabricated in his day. But even without them, his belief that a vacuum could conceivably be made by a finite force allowed him to suggest that this is what is happening in the pumps when the water fails to rise as far as it otherwise should. Water is breaking away from the plunger under the force of its own weight and leaving behind a vacuum.

As in the case of Galileo's observations of pendulums, what we have here is a new point of view changing the significance of an old, well-known observation and aiding the creation of a new problem. Galileo does not do this single-handed, for several generations of attempts to improve pumps were an important factor. And this development doesn't solve the problem at hand for Galileo's account was in fact wrong. The correct explanation awaited Torricelli and his experiment with a mercury barometer. But Torricelli was a pupil of Galileo and his experiment was made possible by the way Galileo isolated this problem.

But we need not restrict ourselves to examining these ways in which aspects of Greek atomism came to influence later thinking and research. One might even argue that since none of these effects involve the fundamental, indivisible nature of basic particles, they ought not be called atomism at all. Nonetheless, the effects of Democritus's complete view of the world are equally clear at this time. Galileo, for example, had suggested that a ball rolling on a smooth, horizontal plane will continue with a constant velocity in a straight line forever. This is very close to what we now call the principle of inertia, but it misses it nonetheless. For it holds only for motion on a horizontal plane.

Nor is this surprising, for one can't get this notion of inertia from experimentation. While balls rolling on horizontal planes do go on for some time compared to those rolling up or rolling down inclines, they do not go on forever because of the inevitable effects of friction. The law, that is, is an idealization and an unlikely one for Galileo to believe in because he is still in many respects an Aristotelian. He has rejected the distinction between natural and violent motion, but horizontal and vertical motion still seem different. They are as different as up and down versus sideways motions understood, importantly, relative to the earth. In an atomistic universe, however, there are no ups and downs. For the elementary corpuscles swimming in an infinite void, there are no directions because an infinite void has no directions.

It was the atomist Descartes who first enunciated this principle in its full generality, and in a way that led Galileo to his limited version. Atomism led

Descartes still further and provided him with an entirely new sort of dynamical problem. This is the problem of impact. Since all changes in the motion of elementary corpuscles are brought about through the impacts or collisions, scientists had to study those impacts and ask, for example, What happens when two elementary particles collide? Or, How is motion transmitted from one to the other? The laws Descartes claimed to find in these collisions were bad and were later corrected in a way that led to the law of the conservation of momentum. But this study of impact was itself new and could not have been found in the scientific literature before. Now, it is almost the pre-eminent dynamical problem—the problem of billiard balls. But it arose not in that form, but rather under the influence of the atomistic view of nature.

There are many more examples of this influence. Besides these and other cases in dynamics, they occur in the study of heat, of light, and of chemistry. Instead of listing these in more detail, let me remark on the extent to which these developments culminate in and are given a new form by the work of Isaac Newton. I have already suggested one sense in which atomism provided a motive and a direction for Newton's research, the sense in which the entire world can be understood as a machine. On this view, again, the universe is like a watch, composed of the same material throughout, and whose parts operate in essentially the same manner everywhere. This conception, as much as watching the apple, lay behind Newton's search for a "universal gravitation" that would describe the same force acting on the moon as well as an apple.

But the adjective "universal" has another meaning which is, I think, even more thoroughly atomistic. This meaning concerns universal gravitation understood as an attraction inversely proportional to the square of the distance between the center of any two bodies. Newton himself, however, was dissatisfied with this formulation. He demanded and then found a proof that if all the little particles within a body acted this way the resulting effect would be as though the entire mass of any body were concentrated at its center.[3] This demand was new. Kepler and others working in this area were quite content to consider forces acting between bodies without so conceiving them as massive points. Newton's insistence that laws of force be understood in this way as acting between corpuscles and therefore also between bodies is therefore new and I think its source lay in the notion of an atomistic world machine. Descartes, whose system was not successful had imposed the same demand, but I know of nothing which parallels it before the popularity of atomism.

Newton's fruitful application of atomism was not however restricted to dynamics. Newton was able to account for Boyle's law on the basis of repulsive forces acting between particles and in optics, including reflection, refraction, and simple diffraction, he was able to make progress on the basis of atomistic ideas. After Newton, everyone was an atomist. But this was a new sort of atomism not the same as that of Democritus or Leucippus. This was not just atoms moving in the void but elementary particles moving under the influence of forces acting between them. This notion of forces acting at a distance was new and radical, as well. For until this time the only thing affecting the motion of elementary particles were impacts. The methodological importance of this was a shift in the aim of physicists: to discover new sorts of forces between bodies and to determine the effects of such forces. Not all, but much of the science of the eighteenth and nineteenth centuries was motivated and directed by this new kind of atomism which emerged from the Newtonian synthesis.

Since Newton's day there have been too many fruitful applications of atomism and too many modifications of atomism to permit our undertaking even a superficial sketch. There is too much to cover and it is largely too technical. I will however reserve for the end of the hour a few more general remarks on the subject of modern atomism. For now, I should like to illustrate in a more detailed way how atomistic ideas can suggest new significances to old data, and of the manner in which atomism itself is modified by the application.

For this purpose one could scarcely find a more central or typical figure than the English chemist John Dalton, whose work at the beginning of the last century brings genuine atomism into chemistry. It is perhaps better to say that Dalton's work shows the fruitfulness of atomism for chemistry, because even before Dalton's time a number of influential chemists believed that substances they studied were built up of atoms. This was an atomism borrowed directly from the physicists of the period. And its use was to explain those properties of natural substances that we should now call physical.

Consider Dalton's pile-of-shot model of matter:[4]

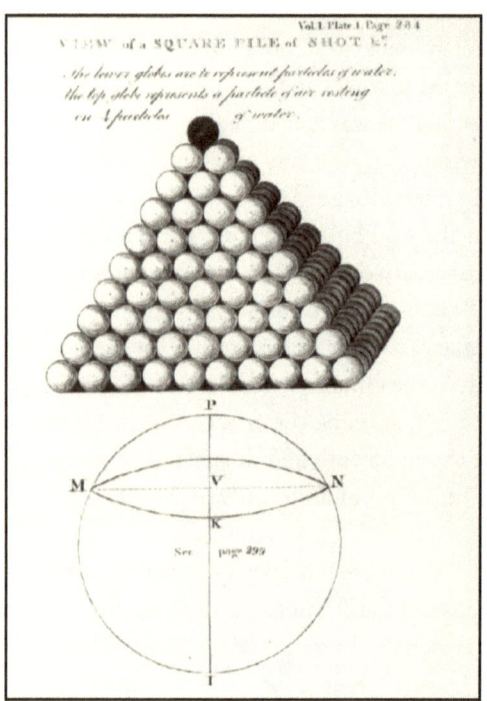

"View of a Square Pile of Shot," from Dalton's essay "On the Absorption of Gases by Water and Other Liquids."

Each shot is an atom and the pile of them fill space, while the individual particles are themselves composed of two substances—the Core, which accounts for the weight and the nature of the substance, and an outer sheath of Caloric, a weightless fluid or jelly in which the particles are set. The Caloric sheath provides the forces between the individual atoms.

This model is useful in several ways. It accounts for the difficulty in compressing matter, but also accounts for the expansion of objects or gases by heat. Caloric, after all, is just heat. This includes the way objects expand in all directions upon heating and it also includes changes of state (as atoms of a solid become a liquid, for example) and the different heat capacities of different substances. These are admittedly not quantitatively useful features of the model, but they illustrate how it reduced and ordered different and seemingly unrelated phenomena. That said, these are not chemical phenomena, either. There is no notion here that these particles enter into reactions with each other. It gives us a way to understand different kinds of substances—Hydrogen particles, Oxygen particles, and Water particles—but this is a static model that does not illuminate chemical changes of state.

These feature of Dalton's thinking are perhaps best explained by the fact that he was not a chemist but a physicist, one who was deeply influenced by Newton's atomism. To be more precise, his work was mostly in meteorology and the study of heat. As a meteorologist, though, he was much concerned with a chemical discovery that he made when he was in his teens. It was the discovery that air is not simple but compound. It consists of two gases, oxygen and nitrogen, plus some water vapor.

This itself raised some serious difficulties that Dalton tried to understand. One, the two gases are different in weight. But why don't they form strata, with oxygen, being heavier, at the bottom? Another problem concerned gas absorption and the puzzle of why some, but not all gases could be absorbed by water. Dalton attempted to solve these puzzles by considering different ways of stacking atoms, different sizes for the caloric sheaths, and different kinds of force laws that might be operating. Through this he convinced himself that no one force law could possibly account for the lack of stratification of mixed gases or absorption phenomena. But he also concluded that a force law could be worked out satisfactorily only if the atoms, with their sheaths, were of different sizes and weights. There is no point in our examining the details of this theorizing, partly because it was bound not to work and was in some ways unusually absurd. The important point was that it directed Dalton to a new research problem: to find the sizes and weights of the ultimate corpuscles in the substances he studied.

It was Dalton's great genius to point out that once certain assumptions were made, this could be done on the basis of existing chemical data. The assumptions in question were that that substances were made of atoms and that all atoms of a given substance are identical. (If they weren't, we'd have two or more kinds of oxygen, for example, which complicates the situation unnecessarily.) On the basis of these assumptions, Dalton then found that one can compute the relative weights and sizes of the fundamental particles. For example, we know that 8 parts by weight of oxygen combines with 1 part by weight of hydrogen to form water. If this combination involves just two atoms, then the weight of oxygen atoms is 8 times the weight of hydrogen atoms. It could of course be a combination of two hydrogen atoms to every oxygen atom, in which case the weight relation would be sixteen to one. But Dalton wisely refused to complicate the situation more than was required and proceeded to determine the weights of other kinds of atoms as well as the relative volumes of the atoms or corpuscles. He was able to do this because knowing that atoms are space filling, and knowing the relative densities and relative weights of the fundamental particles involved, you can discover their relative sizes. In this way he created a list of the relative weights and relative sizes of the atoms. The sizes were different and he convinced himself that his mechanism for understanding the atmosphere would work out. Dalton himself was *delighted*.

Now, if this was all that there was to Dalton's theory, no self-respecting chemist would have paid any attention. The puzzle over atmospheric mixing was not seen as a big problem, and it took very little perspicacity to see that Dalton's theory of forces between particles would not really account for anything. In addition, the notion of atoms uniting with each other to form new kinds of atoms was speculative and did not seem to explain why it was that atoms should unite as they do. Why should it not involve 2 oxygens to one hydrogen, two hydrogens to one oxygen, or 7 to 11, or something else? Dalton's theory provided no answers.

But it turned out that however much Dalton's speculations were useless for solving the problems that interested him, they were of potentially great use for chemists. There was at this time, for example, great debate among chemists about whether substances can combine in any proportion whatsoever. While hydrogen and oxygen show only one proportion, it remained that copper and oxygen seem to combine in almost any proportion you may choose. If Dalton was right, there were only certain proportions in which elements could combine, which meant that other seeming combinations must instead be mixtures of elements. Dalton's research thus introduced a clear criterion for distinguishing chemical from physical combinations of elements and this was of very great importance.

Another striking consequence of Dalton's work for chemists concerned reactions of elements that combine in more than one ratio to form different compounds. Dalton suggested that in such cases one compound must be binary and two must be ternary. In the case of a nitrogen and an oxygen atom, for example, the two can form nitrogen monoxide, NO, or three atoms can combine to form either the ternary nitrous oxide, N_2O, or nitrogen dioxide, NO_2.[5] (The same is true for the carbon oxides, carbon monoxide, carbon dioxide, and dicarbon monoxide.) Thus emerged Dalton's law of multiple proportions, according to which a given amount of one element will combine with weights of a second element which bear to each other simple whole number ratios. Interestingly, this data had been available for years, and in Dalton's time a number of chemists were planning to announce the existence of this regular pattern for a few particular compounds. But Dalton superseded them by showing the mechanism at work and its generality (with the result that many of these other chemists did not even publish their results).

So we see here how a new law entered chemistry, a new guiding principle to be used for all chemical manipulations, which came from a completely irrelevant source. For the gas problems with which Dalton was concerned were not relevant to chemists. And the central parts of this discovery from the point of view of the chemists were mere asides for Dalton. Searching as he was to understand the forces between fundamental atomic particles, the ratios in which elements combined had relatively little significance for Dalton's research.

Just how irrelevant Dalton's considerations were is perhaps shown by an immediate sequel to his discovery—a sequel which actually destroyed its theoretical underpinning. In 1809, just two years after Dalton announced his theory, the French chemist Gay-Lussac discovered another regularity in chemical reactions, the Law of combining volumes. This law states that volumes of gases, for example, combine among themselves in very simple proportions as ratios of whole numbers.

This result was immediately accepted with rejoicing by all chemists, who saw it as a beautiful proof of the atomic nature of their substances and their reactions. But Dalton could not believe it. He doubted not only the generality of the law, but he accused Gay-Lussac of twisting his figures to produce the apparent regularity. This is not as strange as it may seem, because Dalton's insistence upon differences in volumes meant an insistence upon different numbers of particles per unit volume. But the simplest interpretation of Gay-Lussac's result was that there were equal numbers of particles, or at least whole number ratios of numbers of particles, in the volumes of gases. And this would not fit Dalton's data at all. Within two years of his proclamation of his theory it was destroyed. But Dalton held out until the end of his life. As far as these chemists were concerned, however, Dalton's theorizing was not fully destroyed, for they were not committed to his theory of forces or to his computation of sizes of the different elements. They simply rejected the part of Dalton's theory that he held on to while drawing great profit from those parts useful to them.

This does not mean, however, that the chemists were out of difficulties.[6] If you consider, for example, the reaction where oxygen and hydrogen combine to create water, the fact that two volumes of hydrogen combine with one of oxygen seems to mean that the resulting compound is H_2O. But this raises the question of where the extra oxygen atoms come to form the extra volume of steam? There are two ways of solving this problem. One is to suppose that there exist only half as many water particles per unit volume. This is plausible, but it gives a certain arbitrariness in chemical formulae. The same might be true of hydrogen, in which case water is not H_2O but HO.

A better method was suggested by the Italian physicist Avogadro two years after Gay-Lussac's discovery. Avogadro believed that equal numbers of particles exist in equal volumes of gas, but that these particles are not ultimate. They can be broken down in chemical reactions. So when water is formed by combining two volumes of hydrogen and one of oxygen, these substances are dissociating and recombining. This view solves the problem, but it's absurd. For if an atom means "undivided," then it means these atoms of oxygen and hydrogen are not really atoms at all and only more questions must be addressed. If they can be broken in two so as to form water, for example, then why not into three, five, or a hundred fragments? At this point, you may ask, what good is the very notion of atoms?

Because of these confusions, Avogadro's proposal was scarcely even taken seriously for almost fifty years. This is a long story that I can't retrace here, but I would like to note for future reference that this rejection of Avogadro's proposal caused great difficulties. For the proposal effectively linked atomic weights to measurable volumes, thereby establishing a standard for the weights of elements. But without that, different chemists wrote atomic weights and molecular formulas differently. What was the formula for water and what was the weight of oxygen? Different answers caused very serious difficulties for all of chemistry. By midcentury, as the number of known compounds proliferated, the situation became almost unbearable. Different chemists could not read each other's papers without having in hand a table of the atomic weights used by the author in question.

The difficulties were so acute, it almost led chemists to abandon atomic theory itself. As it happened, however, with the atomic theory shaken to its foundations, it became a great deal easier for chemists to eventually relinquish the traditional conception of atoms as indivisible and inviolable and to return, by 1858 and under the leadership of the Italian chemist Cannizzaro, to the modern view which finally provided chemists unanimity in their tables of weights and of formulas. This required, of course, a new sort of atom—one that could be broken; one that did not fill space like a pile of shot; and one without the caloric sheaths that once explained some physical properties so nicely.

To conclude then, we see that these chemical atoms emerging at the end of the nineteenth century were different from any we've met so far. They were not all made of the same kind of stuff, as were Democritus's atoms. They could be split in certain ways during chemical reactions. And they did not have any function in explaining the physical properties of the substances into which they entered. The function of explaining these physical properties were at this time taken over by a different sort of atom developed by the physicist—the atom of the kinetic theory of gases. Here the atoms are very small compared with the space they occupy. And they are in continuous, rapid motion as they move through this vast empty space. Each is like a hard little ball that rebounds with perfect elasticity when it hits another atom.

In fact, the possibility of reconciling these two sorts of atom seemed so remote that there was at the end of the century a widespread rebellion against atomism in general. Some admitted that atoms may be useful devices in terms of which to think. But it was said that science itself had shown that the notion of fundamental particles from which the universe is constructed is absurd. And to have believed in them, we were nearly as bad as the Greeks who indulged in similar kinds of futile speculations. However useful atoms may have been, it was said, the science of the future should purge itself of such notions.

Fortunately, this movement was not successful. In the last twenty-five years—and only then—we've successfully reconstructed an atomism which accounts for both the physical and chemical properties of atoms. It even tells us a good deal about the qualities of atoms in an aggregate. Again, however, we are dealing with atoms of a new kind. This is the planetary atom, with electrons spinning around a hard nucleus which itself can be split into an increasing number of different sorts of particles.

What our notions on the subject will be fifty years from now is almost impossible to predict—but they will be different. The new ideas about fundamental particles will be achieved by the only tools we know how to use, namely our existing and inadequate gained from our examination of the planetary atom. For just as in the earlier reshaping of atoms from those of Leucippus to those of Dalton, it is in the course of applying our old tools to new problems that discoveries will be made which will again reshape our notions of atoms and of the nucleus.[7]

Notes

1. The quotation comes from Epicurus's Letter to Herotodus and appears in *Epicurus: The Extant Remains*, trans. Cyril Bailey (Oxford: Clarendon Press, 1926), 31.

2. Kuhn notes here that the next three paragraphs may be omitted if the lecture is running long.

3. Kuhn neglected to write "the square of" in this formulation of Newton's law of gravitation.

4. Kuhn's outline directs him to draw Dalton's "pile of shot" model on a blackboard. I inserted this illustration from Dalton's essay "On the Absorption of Gases by Water and Other Liquids," *Memoirs of the Literary and Philosophical Society of Manchester, Second Series* 1 (1805), 271–87.

5. Kuhn's outline instructs him to illustrate these reactions for his audience on his blackboard.

6. Kuhn marked this discussion of Avogadro as "omit if necessary."

7. Kuhn noted "read" next to this final section, which I have altered slightly.

Lecture IV

"The Principle of Plenitude": Subtle Fluids and Physical Fields

In the last lecture we examined some aspects of the rise of the atomic or corpuscular philosophy during the seventeenth century, and we noted certain new scientific insights and abstractions which an adherence to this underlying metaphysical scheme had facilitated. Finally, we illustrated with the work of John Dalton the manner in which successive modifications of the atomic philosophy in the period from the publication of Newton's *Principia* until our own day have accompanied the development of new and fruitful scientific theories.

The historical coincidence of the reign of atomic theories with the reign of science has led some philosophers and historians to identify science with the study of the behavior of fundamental particles moved by mathematically describable laws of force. The atomic picture of the world, together with the correlated notion of the world machine running throughout eternity under the dominion of immutable laws, has normally been labeled materialism, and science has been both lauded and vilified for the materialistic view of nature thought to be implicit in its methods and in its results.

I do not wish to involve us now in a discussion of the ethical correlates of science, ancient or modern. But I should like to dwell upon the historical fact that the progress of science has not been inevitably correlated with the atomistic philosophy, that there are other fruitful modes of scientific thought, and that atomistic thought has on some occasions retarded scientific progress in one field of research at a time when it was facilitating investigation in another. In particular I intend to devote the bulk of this evening's lecture to the discussion of an example of the fruitful use in science of an alternate mode of thought.

This alternate manner of regarding the behavior of the external world is loosely described in the title of this evening's lecture by the phrase "subtle fluids." Unfortunately I know of no one definition of a subtle fluid adequate to cover all the occurrences of some such notion in the history of science and philosophy. I shall not therefore even attempt a precise definition, but will be content with a vague description of the idea. The notion will I believe become clear enough as we consider further examples of particular subtle fluids during the course of the lecture. We must, though, understand from the start, that the notion of a subtle fluid is not, like the notion of an atom, correlated with a complete metaphysical picture of the structure of the world. Subtle-fluid thinking may enter into a cosmology, but it is not itself a complete cosmology. On the contrary, except in the most ancient and perhaps in the most recent times, subtle fluids have been applied to the explanation of particular phenomena, of particular limited aspects of nature.

In its most general sense, the concept which I have labeled that of a subtle fluid is the concept of a corporeal or incorporeal substance which resides in matter and is responsible for some or all of the qualitative properties and the effects of that matter. The subtle fluid is a principle which bears qualities. Historically it is very narrowly associated with the medieval concept of a substantial form. It is a thing in the sense that its effect, its behavior, can be described. But it is normally not corporeal in its pure form. You cannot pick up or collect a cubic foot of subtle fluid and perform experiments on it.

Technical variants of the subtle fluid are clearly visible in modern physics, but the notion is better illustrated by reference to commonsense examples of which science has made great use and from which it has now departed. Most everyday thinking about heat, for example, employs the notion of a subtle fluid. Heat is the fluid associated with the quality temperature. The stove puts heat into the pan of water, thereby raising its temperature. When the pan cools, the heat flows from it to the surrounding air or to the cool table surface on which the pan has been set. We shout, "Close the icebox door before the heat gets in." But you can't pick up heat, collect it in a measuring cup, or talk about its shape and color, and the same is true of the popular notion of electricity.[1] It flows in wire, lights our lamps, or charges an electroscope, but you can never handle it in a pure form.

Like the idea of an atom, the notion of a subtle fluid is very old. In one of its important forms, it was the ethereal fluid which filled all space and prohibited the existence of a vacuum. Variants of the space-filling fluid occur as Plato's *anima mundi* and as the *pneuma* of the Stoics. In this last modification, the subtle fluid becomes a vital principle, a form of the soul, and as such it was closely involved with theological controversy during the middle ages. In various guises then, the notion of a subtle fluid can be traced from the Greeks through medieval philosophy and theology into the early science of the fifteenth and sixteenth centuries. And here again the form of the fluid appears as a medium which fills all space. It is, of course, the elimination of this notion of a full universe which paved the way for Galileo's laws of motion and for the discovery and explanation of the barometer.

But having illustrated in previous lectures the manner in which underlying metaphysical notions can in the course of their development facilitate or retard particular scientific insights, I should like this evening to omit consideration of the historical continuity of the development of the idea of a subtle fluid in favor of a more detailed examination of one of its more recent applications. In particular I shall omit all further remarks about that aspect of subtle-fluid thinking referred to in the title of this lecture as the principle of plenitude.

Instead, let us now turn to a consideration of the role of subtle fluids in the development of chemistry during the hundred years between 1670 and 1770. Chemical theory, or perhaps more accurately the intellectual rationalization of chemical practice, was during the first half of the seventeenth century, as during much of its previous history, dominated by subtle-fluid thinking. According to a simplified, unified, and overtly coherent version of this rationalization there were four basic elementary principles: Earth, water, air, and fire. These were considered to be four different substantial forms which could be imposed upon neutral or base matter. And the characteristics, the qualities of that matter, were then determined by the amount of the relative proportions of the various principles residing in it.

The various earths found in nature contained a superabundance of the principle, earth; water, or oils, or liquids generally were composed of base matter whose primary form was determined by an abundance of the element water and whose more important differences were to be accounted for by varying proportions of the other three elements. Chemical changes were explained as processes by which the relative abundance of the four elementary principles were altered. Some chemists believed in addition that it was, under extraordinary circumstances, possible to convert one of these principles into another, that is, to perform a true transmutation. But many other chemists regarded transmutation as a process in which one dominant principle was driven away from raw matter by the action of fire and another principle was substituted for it. Transmutation was, on this more prevalent theory, simply a particularly striking case of a chemical reaction.

The chemistry of four elementary principles which carried with them all the complex qualities of terrestrial bodies was, viewed in retrospect, incredibly simple-minded. And, as a scientific theory, it was virtually useless; in this period the art of chemical practice immeasurably transcended its theoretical substructure. But as a rationalization of crude observations it was far from implausible. Three of the elements, earth, water, and air, corresponded to the three states of aggregation which are still recognized in all scientific literature: the solid, the liquid, and the gaseous state. These are certainly the most striking differences between natural substances. And, if you apply to a naturally occurring substance, particularly to an animal or vegetable product, the most usual tool of seventeenth-century chemical analysis, that is, distillation, you do observe that during the disintegration of the wood or the plant that fire leaps up to the heavens, that smoke is given up which dissipates into air, that tars and liquid juices are squeezed out of the plant or wood being analyzed, and that the final residue is an ash which crumbles into fine dirt or earth. Most natural substances contain all four elements.

Furthermore, you do have in this very elementary and inadequate form of chemical theory the germ of some very important modern scientific notions. At least implicitly, we are confronted in this antique chemistry with the modern idea that the great complexity of naturally occurring objects is produced by the combination in various proportions of a rather small number of elementary substances or elementary principles and that, by the application of appropriate chemical techniques, these complex bodies can be uniquely reanalyzed into the elements which originally gave them birth.

In the second half of the seventeenth century this whole notion of chemistry based upon the analysis of compound bodies into enduring elementary principles came under violent attack, particularly from Robert Boyle, the leading exponent in this period of the application to chemistry of the corpuscular or atomistic philosophy of the day. On balance, the effects of the attack were highly salutary, for they pointed out the glaring inadequacies of previous chemical theory. Boyle, for example, emphasized the tremendous differences between the various substances lumped together as forms of earth, or of water. Again he pointed clearly to the difficulties raised by the fact that small changes in the distillation procedure could produce large changes in the products of the chemical analysis. And he questioned the grounds of the belief that distillation separated a substance into pre-existing parts, because it seemed always impossible to reassemble the products of a distillation in such a manner as to recreate the original substance. Generally he asked what grounds there could be for believing in the validity of the four Aristotelian elements. Perhaps, he said, more elements were needed, perhaps fewer; but in any case the ones in use at this time were, he said, totally useless. And most of his criticisms, which were by no means unsympathetic, were entirely justified.

But Boyle was constructive as well as critical. He called for the application to chemistry of the corpuscular philosophy which had led to so many fruitful results when applied to physical problems. And these positive theoretical contributions, had they been taken more seriously, would have led chemistry into a blind alley, for the corpuscular philosophy as it was developed in the second half of the seventeenth century did not provide a propitious climate for chemical progress.

Boyle, you remember, believed that there was only one form of matter, that this was divided into innumerable small corpuscles, and that most if not all of the properties of substances which existed in nature were to be accounted for not by the nature of the fundamental particles which composed it but by the arrangement of these corpuscles and the manner in which they moved.

This is a perfectly logical point of view. It could perfectly well have been true, but it was not. And because it was not, adherence to it carried potentially disastrous consequences for chemistry. For any chemist who believes that all natural substances are made of the same sorts of particles, and that the differences between these substances are produced by differences in the arrangement of these common particles, must doubt with Boyle the very existence of any such things as chemical elements.

In *The Skeptical Chemist* Boyle provided a classic definition of an element, a definition to which most modern chemists would be glad to subscribe. And some historians who have examined this definition out of context have wondered why it exerted so little influence on the subsequent development of chemistry. But to ask such a question is to misunderstand Boyle's purpose and meaning. The definition is proposed argumentatively and dialectically, and Boyle proceeds from it to a demonstration that one must doubt, on the evidence, whether such a thing as an element exists. Certainly, he says, the empirical evidence for them is extremely limited. And besides, why should there be any such thing? If all the qualities of natural bodies can be accounted for by the motions and arrangements of the particles which make them up, and if all the particles are themselves made of the same sort of stuff, then it should be possible by suitable manipulations to transform any body into any other body. So, there are no permanent elements or principles.

Boyle was not quite this categoric. "I would not say," he wrote,

> that anything can immediately be made of everything . . . yet, since bodies, having but one common matter, can be differenced but by accidents, which seem all of them to be the effects and consequents of local motion, I see not why it should be absurd to think, that . . . by . . . an orderly series of alterations, disposing by degrees the matter to be transmuted, almost anything may at length be made anything.[2]

On this view there are neither elements nor are there unique chemical analyses. Any substance found in nature can in principle be broken down into an arbitrary number of other substances, and these new substances can, if the appropriate techniques are known, be transformed at will into any other substances.

But chemical reactions, unlike nuclear reactions, are not of this sort, and the effect of the point of view was to direct the attention of Boyle and his followers in the corpuscular school to a series of experiments, which, although of great interest in themselves, contributed very little to the progress of chemistry in that or the next century. For Boyle was led to devote himself particularly to the sort of chemical and physical changes in which it appeared that the arrangement of the particles of some one substance is altered by the action of some external agent. Thus he discussed the way in which the white of an egg is altered by the heat of the sitting hen to provide the various tissues which make up the newborn chick. And he considered the way plants transmute the pure water on which they are fed and fabricate from this water their own tissue. Again, on the more strictly chemical level he examined the changes that can be produced by dissolving a substance like camphor in a number of different liquids and then recrystallizing it. And, at this time, these experiments did not contribute appreciably to chemistry.

It is tempting to ascribe to the prevalence of the corpuscular philosophy in the seventeenth century the lag of a hundred years which occurs between the revolution which produced modern physics and the revolution which produced modern chemistry. But the evidence will justify no such far-reaching conclusion. We can, however, point out with some certainty, that the atomistic view which was so major a facet of seventeenth-century thought contributed materially to the physical revolution of that century while providing at best a sterile soil for chemical progress.

Chemistry in any case continued to advance in terms of a revised system of elementary principles. Only a hundred years later, when the nature of a chemical element had been clearly stated and illustrated by a number of examples was the chemist in a position to absorb physical atomism. And in adopting atomism, the chemists changed it by insisting on the radical differences between the fundamental particles of different elements. Atoms ceased to be made from the same sort of stuff.[3]

We cannot here survey the development of chemistry from the time of Boyle to the chemical revolution of Lavoisier, but much of it can be read as an extension of the chemistry of qualitative principles which we have already examined. But these principles were no longer the old principles. Following the biting criticisms made by Boyle and his contemporaries, chemists begin to speak of principles of acidity and principles of causticity or alkalinity. Again we discover in the literature of this period reference to a fossil salt which underlies all the chemical salts found in nature, and we find again the earthy principle, but now much restricted to apply only to the metallic ores. These classifications are still a long way from those of modern chemistry, but they have a decidedly familiar sound: acid, alkali, and salt. Air remains the only principle capable of producing a gas until the time of Lavoisier himself, but an increasing number of essential modifications of the gaseous principle are recognized.

Gradually during the century the number of principles multiplied, and natural substances were grouped more and more into categories recognizable by reproducible experimental criteria. Gradually the gross qualitative criteria like color were abandoned in favor of chemical tests like those we use today to register acidity. As these categories multiplied and stabilized and as skepticism mounted about the possibility of transmuting a substance in one category into a substance in another without the intervention of some third material, the notion of an elementary principle was replaced by the notion of an element. And these elements satisfied exactly the definition provided earlier by Boyle, who had not thought there could be such a thing. In many cases the transition from a chemistry of principles to a chemistry of elements was so gradual that the change was scarcely noticed, either by the chemists themselves or by the historians of chemistry. For the change in terminology involved no change in chemical theory. Once transmutations had been declared impossible, there was no way of distinguishing base matter inhabited by a given chemical principle from the corresponding element.

The mode of transition is particularly clearly seen in an examination of the subtle fluid, phlogiston, which served the eighteenth century as a principle of combustibility and metallicity. First proposed by the German, Johann Becher, in 1669, the new principle was named and given its final form by another German, Stahl, thirty-five years later. No one of the numerous chemical principles which directed eighteenth-century chemical thought was more readily or more generally accepted by the chemists of the period, and, though today it is customary to sneer at the phlogiston theory, it was in its own day an extremely fruitful conceptualization.

At its birth, the subtle fluid phlogiston was little more than a new form for the old elementary principle fire. Phlogiston was the intangible fluid given off whenever a substance burned. Anyone who has ever watched the flame of a candle knows that there must be some such substance given off. Watch the way the flame leaps away from the wick. Something formerly imprisoned in the body of the candle is surely escaping. And any other substance which will burn must be in possession of the same principle, phlogiston.

Perhaps this sounds like the merest superstition, but observe the scientific structure than can be erected on this base. It is well known that if you burn a candle in a confined space it gradually flickers, grows dim, and finally goes out. The air that is left will not support respiration. Apply the standard test, imprison a mouse in this air, and the mouse will die. The air has been spoiled, lost its vitality, because it has been saturated with phlogiston. And once it has been so saturated it will support neither life nor combustion. It is dead air; it has been corrupted.

That the two phenomena, life and combustion, should be so correlated was one triumph for the theory. It was generally known in this period that human beings and animals live by burning combustible fuels, the grains and grasses which they eat. And this combustion was thought to be the source of the body's heat. What more natural than that the phlogiston given off in the process of internal combustion should saturate the air which we inhale and that this air should be exhaled in its dephlogisticated or spoiled form, just as though a candle had been burned in it.

The air thus functioned as a sponge for phlogiston. Its natural function was to absorb the foul effluvium given off by fires and animal respiration. And when saturated with phlogiston, the physical characteristics of the air were changed. Normal air is a highly elastic fluid. It can be compressed, and it will then regain its own volume, just as a spring will. But if a candle is burned in air until it goes out, the volume of the air is permanently decreased. The spring of the air, like that of a metal, has been spoiled in the process.

So far we have seen the elementary principle phlogiston as a principle common to all combustible substances and as a subtle fluid which poisons the air that absorbs it. But the utility of the principle by no means ends here, and it is in its third role that phlogiston achieves its historically most important function. Charcoal, as you know, is a substance which burns particularly completely. It leaves only the smallest residue of ash. So it must follow that charcoal itself is almost pure phlogiston. It is in fact a compound composed predominantly of phlogiston plus a small amount of ash.

Now for centuries it had been common metallurgical practice to reduce earths, that is, metallic ores, to metals by heating them in the presence of a large amount of charcoal. This process could now be explained in terms of the phlogiston theory. In the reduction of an ore to a metal, a process which uses up a large amount of charcoal, the elementary principle phlogiston was transferred from the charcoal to the earth. And the addition of phlogiston converted the earth to a metal.

That all metals should contain a common principle was itself to be expected. For all the known metals were very similar in appearance: they were all shiny, they were all heavy, they were all malleable, they were all excellent conductors of heat. On the other hand the ores from which they were derived were very widely different in appearance. Some were red, some were black, some yellow and gray; some were hard, heavy, rock-like, while others were fine powders. What more natural than that the entry of some common principle into these very different elementary earths should produce a greater unanimity of properties, should make them all metallic? And so the single fluid phlogiston acquired a third function. It became the principle of metallicity.

The phlogiston theory thus provided a conceptual framework within which the oxidation and reduction of metals could be studied in detail, and historically this proved its most important role. For these reactions were particularly suited to provide fruitful clues to the research chemist. They were, in the first place, simple reactions, although this could not have been recognized until later. The metals and their earths or oxides are chemically simple compared to the animal and vegetable matters to which so much previous attention had been devoted. Again the substances involved in these reactions could be readily obtained in a pure form, so that the reactions in which they were involved were more nearly reproducible than most of the reactions studied previously. And finally, these reactions were reversible. The metallic ores could be reduced with phlogiston from charcoal to yield the pure metals, and these metals could then be returned to their ores by heating them in the presence of air. If the air in which they were heated was confined, it was observed to lose in volume and to be spoiled by the phlogiston emitted when the metal rusted. And so the cycle could be repeated again and again, and its mechanism could be studied in detail.

In short, the phlogiston theory was in its own day an eminently useful theory based on sound observation. It was economical in the sense that it permitted a large number of phenomena formerly thought to be distinct, to be treated as the effects of a single natural principle, and it thus measurably reduced the conceptual complexity of a portion of chemistry. And it was scientifically fruitful. For it directed attention to the study of metallic oxides by which, as it happened, great progress could be made, and it suggested new sets of problems to eighteenth-century chemists in the investigation of which they made valuable additional discoveries.

For example, since both respiration and combustion were known to vitiate the air necessary for life by dephlogisticating it, the theory naturally suggested the question: what agency is constantly purifying the air? It was known that the total supply of air was limited, and yet there was no evidence that it was gradually deteriorating. The English chemist, Joseph Priestley, searching for a means to dephlogisticate the air discovered, after a long series of experiments, that growing a plant in foul air would purify it, and thus started a whole new chain of investigations which by the beginning of the nineteenth century had led to an almost modern understanding of the elements of photosynthesis and plant nutrition. Again, the Swedish chemist Scheele discovered oxygen by a chain of reasoning based upon the phlogiston theory.

Furthermore, the theory adapted itself to new discoveries made by other means. Priestley, who discovered oxygen independently of Scheele and who noted very quickly that it supported combustion even better than ordinary air, labeled it dephlogisticated air, air which was even freer from phlogiston and therefore more capable of sustaining fire. When hydrogen was identified as an apparently new sort of gas, it was immediately taken to be pure phlogiston, for it would burn in air and it would reduce metallic ores even without the collaboration of charcoal.

I do not know what else can be demanded of any scientific theory. The phlogiston theory unified conceptually a group of apparently disparate phenomena; it suggested new experiments which in turn in the investigation of which they made valuable additional discoveries, and these new discoveries could be handled within the theory; and it explained very neatly a variety of discoveries made without the aid of the theory. Nevertheless, the theory was wrong, and the manner of its overthrow deserves attention, for it illustrates many facets of the mechanism by which science advances.

It had been one of the great strengths of the phlogiston theory at the beginning of the eighteenth century that all or most of the gases then known could be understood as modifications of common air by varying degrees of phlogistication. For at that time very few distinct gases were recognized. Almost no techniques for collecting gases were known—the pneumatic trough itself was a product of the eighteenth century—and gasses were more normally tested only for their "goodness," their ability to support combustion and respiration.

But during the course of the century the technique of collecting gases over water and mercury was introduced and with it came several very much improved tests which made it possible to distinguish between gases that had been previously thought identical. Some of the gases thus distinguished fitted the framework provided by the phlogiston theory nicely—oxygen and hydrogen, for example—but others could be fitted to the theory only with the greatest difficulty. Carbon dioxide and nitrogen ought, for example, to have been identical, for one was a product of the combustion of charcoal and the other was left as a residue when a metal was burned in a limited amount of air. Both should have been phlogisticated air, and since both were transparent and inert, they were easily confused. But they could be distinguished and they were during the course of the century. There were a variety of ways of handling the problem presented by the distinction. Some chemists just managed to ignore the distinction or to blame it on some additional character of the experiment. Others decided that carbon dioxide was a gas entirely distinct from air, that nitrogen alone deserved the title of phlogisticated air. Still others managed to preserve a consistent theory by saying that while nitrogen was pure phlogisticated air, carbon dioxide was phlogisticated air plus water. But there was no unanimity in the matter.

Another grave difficulty which the phlogiston theory faced increasingly during the course of the eighteenth century derived from the observation that metals actually gain in weight when they give up their phlogiston and become earths. In this reaction the loss of the subtle fluid was invariably accompanied by a gain in weight, and it is this apparent incongruity which makes it so difficult for modern critics to understand how anyone could ever have taken the phlogiston theory seriously. But this again is a misunderstanding, a part of the mythology of science which obscures the nature of research.

The fact that metals gain in weight when they lose phlogiston or when they rust had been known for centuries before the phlogiston theory was even proposed. Yet it was not initially considered to present a difficulty either for chemistry or for the phlogiston theory. For, in the first place, the change in weight was quite small. It was normally less than twenty percent, and was not nearly so striking as the change of color or the change of texture. It was an incidental, not a primary, characteristic of the reaction, and it could be ignored, at least until the more important aspects of the phenomenon were better understood.

Today, of course, even the smallest changes in weight are considered of vastly greater significance than immense changes in qualities like color. And the fact that, until the middle of the eighteenth century, a change in weight could be ignored while a change in color seemed important is symptomatic of a basic difference in the attitudes toward weight of seventeenth- and eighteenth-century chemists.

To the seventeenth-century chemist, weight, like color and texture, was a *quality*. Therefore, like color, it could be imposed upon base matter, and, like color, it could change during a chemical reaction. And this change of weight did not violate the principle of conservation of matter. For matter itself, pure base matter devoid of qualities, was conserved in all chemical reactions. That had been a basic principle since antiquity. But that matter and weight were uniquely associated, that the same amount of matter must have the same weight in all its possible modifications was explicitly denied, both in antiquity and in much of the seventeenth century. There was and is no *a priori* reason why the conservation of matter should imply the conservation of weight.

It was actually the work of Isaac Newton that made weight an intrinsic or primary quality of matter, that made weight a measure of mass, and it was largely through the spread of Newton's influence in the eighteenth century that the weight changes which occur during chemical reactions achieved more than incidental importance. And since it achieved importance its conservation was taken for granted. The fundamental principle of all modern chemistry, the principle that the weight of the substances entering into a chemical reaction must be equal to the weight of the products, was first enunciated by Lavoisier one hundred years after the publication of Newton's *Principia*. And Lavoisier enunciated it, not as a principle derived from experiment but as an axiom, which by the end of the eighteenth century all chemists were prepared to take for granted—once it was pointed out to them.

After Lavoisier, the study of weight changes became a fundamental tool in all chemical analysis. But between Newton's time and Lavoisier's, the attitude of most chemists toward weight relations was decidedly equivocal. Some of them did not consider the gain in weight when phlogiston was emitted to be important. Others held that fire particles entering the metal during its oxidation would account for the change. Still others believed that air was absorbed by the porous earth formed during the oxidation process and that this would account for the change. Again there was no unanimity.

Now the interesting thing about all these difficulties with phlogiston theory (and I've mentioned only a few) is that they did not lead to the abandonment of the theory. In the 1760s and 1770s the theory became more and more complex and more cumbersome. New experiments called for new modifications and distortions of the theory, and though individual chemists adopted appropriate distortions to fit the theories to the new facts, there was a complete lack of agreement as to the particular manner in which the theory ought to be modified for a particular application. As a result, by 1775, when the theory was first publicly attacked, there were really a number of phlogiston theories. Everyone believed the phlogiston theory, everyone used it, but no one could agree on just what it was. And the overthrow of the theory proceeded not from this apparently negative experimental evidence but from a bright idea advanced by a man new to the field on the basis of his repetition of some experiments which had already been performed by others.

Antoine Lavoisier, frequently known as the father of modern chemistry, appears first to have turned his attention to chemistry shortly before 1770. Among his first experiments were two involving the combustion of phosphorus and of sulfur in limited amounts of air. These turn out to have been particularly propitious choices, for in contrast to the experiments on the combustion of metals, sulfur and phosphorus in combustion display weight increases of over 100 percent and produce large and immediate contractions in the volume of the air in which they are burned. Lavoisier found these experiments tremendously suggestive and he leapt to the conclusion that the process of combustion is one in which part of the air is absorbed by the substance which is burned. Both the volume reduction and the weight increase could, he thought, be accounted for in this manner. And he was so convinced of the truth of his explanation and its applicability to all processes formerly understood with the aid of the subtle fluid phlogiston, that he immediately deposited a sealed note describing his conclusions with the Secretary of the *Academie des Sciences*. Then and only then did he proceed to a careful examination of the other combustion experiments on which earlier chemists had based the phlogiston theory.

All of these experiments were, he found, compatible with his original idea. When appropriately performed all combustions reduced the volume of air and yielded a product weighing more than the original starting substance. But this was not much improvement on the phlogiston theory. It explained to be sure the increase in weight, but that could be handled although in a more complicated manner with the phlogiston theory too. And the new theory failed to explain many things which the older theory had made very clear. Why were some substances combustible and others not? Again, why, if air enters into metal during combustion, does the process stop before the air is exhausted? And why is the residue incapable of supporting combustion or respiration? Or why do the metals show such uniform behavior if they are more different in composition than the earths? Lavoisier's theory in this form had advantages only for a man who, like Lavoisier, thought that achieving constant weight relations was the chemist's first responsibility.

Actually the suggestion that air is absorbed or "fixed" during combustion was an old one. It had been advanced by several English chemists during the seventeenth century, and it had been ignored or rejected in favor of the phlogiston theory. The weight relationships were simply not that important. And the reasons for the ultimate success of Lavoisier's theory lay partly in the new attitude taken toward weight in the eighteenth century and partly in an improvement which he was able to give to the theory.

This improvement was the recognition that air is not a simple substance but a mixture of two distinct gases, oxygen and nitrogen. And the experimental ground of this recognition, which had not been a part of Lavoisier's first statement of his new theory, was provided again by an experiment first performed and differently interpreted by a believer in the phlogiston theory, Joseph Priestley.

Priestley discovered that red oxide of mercury, unlike other metallic earths, could be turned into a metal at moderate heats without the intervention of charcoal, and that in this process a gas was liberated which supported combustion better than normal air. He called the gas dephlogisticated air, and thought of it as completely freed of phlogiston. For Priestley the discovery represented a new triumph for the theory.

Lavoisier repeated this experiment with a measured amount of the red oxide. He carefully collected and measured both the gas and the mercury produced by the decomposition. Then he heated the mercury in an atmosphere consisting entirely of the new gas. The red oxide or earth was formed again, and this time there was no residue of gas. Priestley's dephlogisticated air would support combustion until it was entirely exhausted.

This gave Lavoisier the modern answer—the new gas was oxygen, and it was part, but only part, of normal atmospheric air. Normal air was a mixture, and only the oxygen portion of it would support combustion. It was this gas which was absorbed during combustion, and it was the absorption of this gas that accounted for the reduction in volume of the air and for the increase in weight of the resulting earth. The earths were compounds, the metal elementary.

The new theory was a good theory, produced by one of the great geniuses science has known. A modern student comparing it with the phlogiston theory immediately recognizes the greater simplicity of the new scheme. But it was not immediately accepted by Lavoisier's contemporaries. The phlogiston theory could be and was revised to account for all the phenomena explained by the new theory. The revision was cumbersome, but quite adequate to fit the facts. And for its adherents it had two great advantages: it explained combustibility, why things burn, and it explained the common appearances of the metals. Chemists who accepted the oxygen theory had to conclude that the explanation of these phenomena was no concern of chemistry's, and modern investigations have shown them to be wrong, for today we can account for these properties.

During the twenty-five years between Lavoisier's announcement of the new theory and the end of the century more and more chemists switched their allegiance from phlogiston to oxygen. By the end of the century Joseph Priestley, whose experimental discoveries had done so much to create the oxygen theory, was almost the only major chemist to oppose it. And this opposition he maintained until his death.

So by the nineteenth century, chemists had finally readjusted to the large number of new experimental discoveries made during the preceding hundred years. But even this readjustment, modern as it sounds to us, had by no means banished subtle fluids or the chemical principles from an active role in chemical research. For Lavoisier, oxygen itself was still a chemical principle; it was the principle of acidity. The presence of this principle oxygen gave to a liquid its acid properties. This was certainly reasonable, for most acids do contain oxygen; but it was also quite wrong, as the subsequent analysis of hydrochloric acid was to show. Nitrogen, Lavoisier suggested, was probably also a principle, the principle of alkalinity. And these principles, nitrogen and oxygen, were not the gases that we call by these names today. On the contrary, the gases were themselves compounds of the principles nitrogen and oxygen with the subtle fluid caloric.

For it was really caloric, not oxygen, that replaced phlogiston during the first half of the nineteenth century. Caloric was the substantial form of heat, and perhaps also of light. It was this fluid, you remember, which held the small particles of elementary substances apart and which, by its emission during chemical reactions, accounted for the heat and the light evolved.

Just as a major virtue of the fluid phlogiston had been its ability to unify under one conceptual scheme a group of apparently disparate phenomena, so it was the virtue of caloric to permit the explanation of phenomena as divergent as the expansion of matter when heated, the emission of heat in chemical reactions, the different capacity of different substances to absorb heat, and the common physical properties of the known gases.

Phlogiston and caloric are but the beginning of a long list of subtle fluids which have played an important and fruitful role in the development of science. In the seventeenth century, the actions of magnets had been explained in this manner. Newton, despite his general adherence to the corpuscular philosophy had employed a subtle fluid filling all matter in his mathematical explanation of the colored rings formed by thin films, and he had suggested that the same fluid was responsible for the observed effects of refraction and reflection. For many years he had tried unsuccessfully to describe a fluid whose properties would account for the action of gravity, for in spite of the impression to the contrary, he was never satisfied with the notion of action at a distance.[4] And for many years he sought a subtle-fluid explanation for it.[5]

In the eighteenth century, electricity and heat join the chemical principles in the list of subtle fluids guiding the direction of scientific research, and in the nineteenth century many of these reappear in modified form. Subtle fluids vanish from chemistry entirely in this period and are replaced completely by the chemical elements. With this transition chemistry is, as we pointed out in the last lecture, in a position to profit from atomism; but this of course was a new sort of atomism, an atomism which admitted many different sorts of fundamental corpuscles.

Heat had a somewhat similar history. During the reign of philosophical atomism in the seventeenth century, it was normally taken to be a mode of motion. The corpuscular philosophers, led by Bacon and Boyle, examined the sorts of agents which could produce heat. In particular they examined the heat generated by friction and by percussion, for these were obvious examples of the transmission of heat as motion, and they were taken to be fundamental. This portion of seventeenth-century research on heat was fruitless. In the eighteenth century, heat was studied as an entity which flowed from body to body, as a subtle fluid, caloric. And these investigations resulted in the discovery of specific and latent heat. But in the nineteenth and twentieth centuries, heat was *successfully* treated as a product of the motion of fundamental particles, and this subtle fluid has been abandoned.

But subtle fluids, though modified again and again, are not always abandoned in favor of atomistic hypotheses. The modern concepts of light and of the electric and magnetic field are products of a gradual evolution of subtle-fluid thinking. In their present highly mathematical form, these concepts have lost all or most of their resemblance to the weightless fluid, inactive but tangible, from which they have grown. Perhaps the name "subtle fluid" is now as serious a misnomer as the name "atom." But whatever the name, these modern descendants of the subtle fluids remain conceptual entities, distributed in space, and active agents in the generation of scientific data.

Notes

1. Kuhn inserted by hand "the popular notion" here, perhaps because he realized that in some senses electricity can be stored in a jar, as illustrated by electroscopes and the Leyden Jar, a simple kind of capacitor, which he discusses later (see note 5).

2. This quotation comes from Boyle's essay "The Origins of Forms and Qualities According to the Corpuscular Philosophy." See *Selected Philosophical Papers of Robert Boyle*, M.A. Steward, ed. (Indianapolis: Hackett, 1991), 49–50. Kuhn mistook Boyle's "local motion" for "locomotion."

3. At this point, Kuhn noted but later crossed out by hand: "Qualify discussion of substantial form to indicate that there are also some qualities associated by accident, but that these are the unimportant ones. Thus two bodies inhabited by the same substantial form need not be entirely identical."

4. Here Kuhn typed two parenthetical notes, which he later crossed out: "Get caloric and light into the list of Lavoisier's elements" and "Modify the quality-bearing aspect of your original definition of subtle fluids."

5. What follows from this point forward is a shorter version of the lecture's conclusion that Kuhn appears to have typed subsequent to completing an original, considerably longer version of the conclusion. The original, longer conclusion reads as follows:

> In the eighteenth century, electricity and heat join the chemical principles in the list of subtle fluids guiding the direction of scientific research, and in the nineteenth century many of these reappear, modified into both the elastic ether by which light was supposed for a long time to be transmitted and into the lines of force which describe the properties of the electric and magnetic fields.
>
> We are certainly not through with subtle fluids today. In fact, the variety of different subtle fluids which have played distinctive roles in the progress of science is so vast that I cannot resist one last brief example of the way in which the concept of a subtle fluid can simultaneously guide and be modified by experimentation. At the beginning of the eighteenth century the attraction of a glass

rod briskly rubbed with cloth for dust or minute particles of paper in its vicinity was generally accounted for by the presence of a subtle effluvium normally contained in the glass and permanently anchored to it. The cloth rubbing the glass excited the effluvium and caused it to stream out into the atmosphere where it attracted light neighboring particles. The reality of the effluvium could scarcely be doubted. If the rod were held close to the face a tickling sensation like that produced by cobwebs was clearly distinguished. In 1731 an obscure English pensioner named Stephen Gray performed some electrical experiments with a large, hollow glass tube in whose ends he had inserted a cork. To his surprise, he found that rubbing the tube not only excited the glass but that finally the corks themselves gave evidence of emitting the electrical effluvium. By further experimentation, Gray showed that these effluvia could be transmitted over quite large distances. Clearly then, it was not anchored to the glass but was in fact a fluid originally resident in the glass but which could when excited be made to flow away from it. And this fact was also taken to explain the previous observation that the rubbed glass rod does not maintain its electrical properties indefinitely if it is surrounded by air. Fourteen years later, Pieter van Musschenbroek of Leyden tried the law of the electric effluvium in the glass rod or any other charged body to the air. What more reasonable than to surround the charged body not by air but by some substance known to be a very poor conductor? Accordingly, Musschenbroek enclosed water to be charged inside a glass vial and ran a wire through the cork of the vial into the water to serve as the medium of transmission of the original charge. All went well—the water was charged and a friend Andreas Cunaeus approached in order to withdraw the wire from the cork so that no path would be left for the electricity to escape. On touching the vial and the wire simultaneously he received a tremendous shock. Musschenbroek tried the same thing and was even more strongly affected. It was like lightning. No such artificially generated shock had ever been experienced before. A manner of accumulating, of storing up the electrical fluid had been discovered. This was the famous Leyden Jar. Within a year scientists all over Europe were experimenting with the new device. One of these, William Watson, and English Apothecary, noticed that when he grasped the vial with one hand and the wire with the other, the shock appeared to affect "no other part of his body but his arm and his breast" (Whittaker, page 42). This in turn suggested that the actual shock was caused by the transference of the ethereal electric fluid from the wire which was grasped in one hand to the vial which he held in the other. And this is fun. Fluid follows the easiest, or as he thought, the shortest path. With this hint and on the basis of further experiment he elaborated existing theories of the electrical fluid in a manner which sounds almost modern. There is, he suggests, in the world a certain total amount of the electrical fluid, and this is normally distributed in all bodies with equal density. When a body is charged an excess of this fluid is transferred to it from some other body and the ability of the charged body to attract the particles of dust or paper grows with the amount of excess placed in it. If two bodies containing different densities of the electrical fluid are connected, the fluid will flow from the one of greater to the one of lesser density until the two densities become equal. The fluid itself is neither created nor destroyed. With minor modifications this is the theory that you and I learned in high school. (Add some concluding remarks here depending on how it shapes up when you go over it.)

Kuhn left blank spaces for each appearance of "Musschenbroek" and for Musschenbroek's friend and student, Andreas Cunaeus. His reference to Whittaker is likely to

Edmund Whittaker's *A History of the Theories of Aether and Electricity from the Age of Descartes to the Close of the Nineteenth Century* (London: Longmans, Green, and Co., 1910) in which these accounts of Musschenbroek and others appear on pp. 41–42.

Lecture V

Evidence and Explanation

In the first lecture of this series, I advanced, at random, a number of reasons for rejecting the description of the scientist as a man who proceeds without preconception to make dispassionate observations about the world in order that he may discover invariable sequences. And during the three lectures which followed, we examined selected incidents from the history of physical science in an effort to provide material from which this point of view might be documented and from which a more adequate picture of the scientific activity might be reconstructed. It is to this discussion of the role of prejudice and preconception in science that I should now like to return.

But in beginning this second stage of our lecture series it may be better to adopt a slight modification of vocabulary. In the first lecture I called the elements whose roles in science we now wish to examine preconceptions and prejudices, and this choice was not made without a purpose. At the time I wished particularly to emphasize that these elements were, in the first place, in existence prior to the beginning of active research and that they were not normally drawn from evidence relevant to the particular investigation at hand. Second, this choice of terms served to emphasize that these elements, or preconceptions, or prejudices, were frequently unconscious; that they were not normally the result of deliberate rational consideration of the particular problem with which the scientist was engaged.

Finally I hoped that the choice of the word prejudice would suggest that the elements under discussion were individual and alterable. For although I believe they are normally gained as results of training, scientific and social, and that they are therefore characteristics of professional groups for many years at a time, it is equally true that particularly productive steps in physical science are frequently or perhaps invariably associated with the application of a new and individually rooted prejudice to an old problem. Or, to put the same point differently, I wished to emphasize that although these prejudices legislated for experience like the mental categories proposed by Kant, they were not, like Kantian categories, necessary or *a priori*.

But the words preconception and prejudice are negatively colored. In their normal usage they imply an absence of intellectual activity and a regression from rationality. Since I now wish to discuss the role of such elements in science, an activity which I take to be intellectual and rational in the extreme, it might be best to admit them to a more constructive function and to call these elements the points of view of the active scientist or the principles which orient his perceptions and his judgments about the phenomenal world.[1]

That there are such points of view underlying and contributing to all research can, I trust, no longer be doubted. We have examined numbers of them; and we have seen them acting both as sources of scientific inspiration and as blocks to scientific progress. In the third lecture, for example, we noted some of the new physical insight and new laws whose development had been facilitated by an adherence to Greek atomism. We saw, among other things, that the general principle of inertia was for an atomist a reasonable generalization of Galileo's statement about the infinite motion of a ball rolling on a horizontal plane, but that it was not a reasonable generalization for an Aristotelian. And we remarked on the fruitful new problem, the problem of impact or collision, which an adherence to an atomistic point of view had suggested to the philosopher Descartes. But, examining the other side of the coin, we discovered in our last lecture that, for all its suggestiveness to physicists, the corpuscular philosophy had been a real block to chemical progress in the seventeenth century.

We could have duplicated this example if we had studied the development of the science of heat. In the seventeenth century the corpuscular philosophers believed that heat was, like other qualities, a mode of motion. Accordingly, they took as their primary subject of study the generation of heat by friction and percussion. These were the obvious cases of the transmission and generation of heat by motion. But these were, in this period, fruitless experiments. Important theoretical developments in the study of heat waited until the eighteenth century when, under the leadership of Joseph Black, scientists studied heat as a fluid which could be stored in bodies and transmitted without loss from one body to another. And from this viewpoint proceeded a different series of experiments which led to the discovery of specific and latent heat and to the development of calorimetric techniques.

These are, of course, very restricted examples of a phenomenon which we have already observed to be far broader in it scope. A more complete study of the role of points of view in the progress of physical science would require a detailed taxonomic or classificatory study of the types of orientations which have played important roles in the development of physical science. We would have to study, in addition, the logical and psychological sources of such viewpoints, and we would need a more detailed account of the effects which these orientations have upon scientific thought.[2] But such study must await further research. Here I can only suggest certain preliminary divisions and generalizations which seem useful in the examination of scientific points of view.

We have, for example, dealt repeatedly with what might be called cosmological orientation, or cosmological prejudices. These may be described as implicit or explicit views about the structure of the universe. Is the universe finite or is it infinite? Does the universe have a fixed center, and are their preferred directions in it? Or, is space everywhere homogeneous—as it was for the atomists?[3] Again, is the universe made up of atoms? Is the perceptual flux to be accounted for by changes in the positions and motions of these ultimate particles, or are there quality-bearing principles? Are there subtle fluids?

We have examined numerous effects of such cosmological viewpoints during the preceding lectures. But let me illustrate once more, for it will eliminate a simplification which we adopted in the last lecture. You will remember that I there attributed the identification of weight with quantity of matter to Sir Isaac Newton. It was this identification which finally provided so large a problem for the phlogiston theory in the eighteenth century. But the controversy about weight is older than Newton, and Newton's enunciation is in part derivative.

If, as both Aristotle and Descartes believed, the universe is necessarily full, or if to possess extension is necessarily the same as to be material, then the true measure of matter or of mass is not weight but volume or extension. Weight is, in this case, a secondary quality, a product of position and motion and impact. Thus weight need not be conserved, it is of secondary scientific importance, and the meter-stick rather than the balance is the pre-eminent tool of science. And indeed the metaphysical debate about the fullness of the universe is a relevant portion of the intellectual background for the Newtonian annunciation of the proportionality of mass, quantity of matter, and weight.

But not all of the underlying viewpoints which we are considering can appropriately be described as cosmological. Certainly we have observed other sorts of orientations which may but need not be grounded in pre-existing cosmological beliefs. And among these is a type which we may at least crudely describe as metaphorical viewpoints, as tendencies to see certain different sorts of behavior as similar, or as mutually revealing. Indeed all science is dependent upon metaphor though normally the metaphor is implicit and unrecognized. In the first lecture, when we discussed the hypothetical search for a law of the rostrum we assumed from the start that this law when discovered would be applicable to all rostra, or at least all rostra supplied to the same specifications by the same manufacturer. Without this assumption we should not even have started our investigation. But it was an assumption, and it was a metaphor. For it took for granted a similarity as regards the end-product of our investigation of two objects which were not only philosophically distinct but also physically different.

Perhaps you find this a trivial example, but we have been concerned in recent lectures with logically identical examples which do not have the appearance of triviality. Remember, for example, the consequences to dynamics of the reorientation in which motion ceased to be regarded as a change of state and was instead grasped conceptually as a state. Or consider the consequences of the conceptual unification of linear and circular motion. It was this unification, you'll remember, which gave the pendulum a new significance and which led both to new observations and to new quantitative laws. And we observed similar metaphorical problems in our study of chemistry. Were sulfur and gold, because of their common color, to be correlated as bearers of a common principle, or was gold to be listed with the metals because of its lustre and its texture?

There is a third sort of orientation which appears less clearly in the particular historical examples which we have already covered, but which is nevertheless of supreme importance. It may be described as a point of view regarding the sort of questions which scientists may legitimately ask and the sorts of answers which they may legitimately accept. In the last lecture, for example, we observed that the transition from the phlogiston theory to the oxygen theory appeared, quite erroneously, to necessitate the elimination from chemistry of the question, why do things burn? And we noted more generally that nineteenth-century chemistry, by eliminating the qualitative principles which had dominated the field during the preceding century, eliminated as objects of scientific investigation many of the qualitative properties of chemical substances. The quality-bearing principles, like phlogiston, were then labeled occult qualities, qualities without scientific explanatory power, and the questions which their existence had previously answered were dismissed from science. But today we have resurrected these problems and found answers to them.

Our ability to label and dismiss as occult older forms of scientific conceptualization has led to a great deal of misunderstanding of the history of science. And it has now led to many overly narrow definitions of the nature of scientific problems. We scorn as unscientific Aristotle's pronouncement that the planets must travel in orbits compounded of circles, because it elevates the circle to a unique position among geometric curves, and we consider that this gives the circle an occult property. Yet we applaud Einstein's pronouncement that the planets must necessarily travel along geodesics. And I think that no difference between the logical forms of these two laws can be found. We laugh at phlogiston as a weightless fluid which cannot be isolated, yet we speak of electromagnetic radiation which, like phlogiston, can be known only through its effects.[4] I know of no criterion of occultness which will show that we have reduced the number of such conceptualizations in science or that we could profit by doing so.

Our actual criteria of occultness are prejudices of our own scientific generation. They lead to far too narrow a notion of the appropriate problem structure of science. We once, for example, banished with loud huzzas all teleological problems from physics, and, we thought, from science, but we have recently learned from the physiologists how fruitful research on teleological questions can be.

I do not believe that the preceding list of types of orientations or points of view is by any means exhaustive. Certainly there are other sorts, and I suspect that a better list would blur many of the distinctions which I have made above. But we have already proceeded far enough to note many of the functions performed by these orientations, and it is these that I should like now to re-emphasize.

These orientations are conceptual frameworks which suggest the aspect of nature which the scientist ought to investigate. They provide his problems, and changes in orientation produce changes in these problems. In addition these viewpoints dictate to a large extent the sorts of experiments which may be performed in order to discover answers to these questions. And by establishing metaphorical connections within the phenomenal world they provide boundaries within which the scientist must search for his regularities or laws.

But they do far more than this. For as simultaneous directives to problems and experiments they already contain implicitly a major portion of the generalized conclusions which the experimenter will draw from his limited and concrete results. They point toward the particular idealization which the scientist will find illustrated in his experimental data. And again, as we illustrated with Galileo's laws, or with Descartes's derivation of the principle of inertia, or with Dalton's discovery of multiple proportions, changes in the orientation will produce changes in the laws derived from existing data.

These points of view, then, provide limitations upon the form in which our laws may be cast. They thus narrow the gap between the concrete data which can always be interpreted in an infinity of ways and the particular law which in fact we derive from those data. And in doing this they enable us to evaluate our data, for though the gap is rarely so wide as we observed it to be in the case of Galileo, no experiment conforms exactly to the law derived from it. There is always a divergence between the results of scientific experiment and the predictions of scientific law, and our orientations, which in this case appear quite literally as prejudices, provide[5] the ground on which we evaluate the significance of the deviation of the data from the law. Thus Galileo's law was preserved although it diverged tremendously from experiment, for it was the only simple qualitatively correct law which corresponded with the seventeenth-century understanding of what motion was. But Dalton rejected Gay-Lussac's law, although it corresponded very closely with experiment, for if the correspondence was taken to be more than fortuitous it would have destroyed the conceptual basis and the utility of atomism which Dalton had introduced to chemistry.

These orientations therefore appear as vague and qualitative predispositions to more exact and frequently quantitative scientific laws. They suggest problems; they suggest the sorts of evidence relevant for the solutions of these problems; and they suggest the mode in which the answers to these problems must be cast. They are, if you will, predispositions to certain sorts of explanations. But they are not just predispositions toward explanations or toward laws; they are equally predispositions toward evidence, toward facts. They direct our attention to particular aspects of the phenomenal world, and they suppress other aspects. But the sort of law and the sort of fact upon which the law can be based are contained in embryo in the pre-existing scientific orientation.

This suggests that scientific research is inherently circular, that it does not proceed from experimental facts to theories, but that facts and theories are provided together, in a more or less inchoate form, by scientific orientations. On this view the experiment may add greatly to the precision and the scope of the law implicit in the point of view which suggested the experiment. But it will not itself provide a law radically different from that implicit in the point of view from which it derived. If the experiment diverges too far from expectation, and this rarely occurs, it may inhibit or infirm the orientation which produced it, but it will not itself produce a different sort of law. And this indicates that the quest for physical theory may be well described as an attempt with the aid of experiment to apply a given orientation or point of view to the perceptual world. To this attempt experiments lend precision and detail, but they are not in themselves creative.[6]

I believe that this picture of science is clearly confirmed by an examination of the history of science, and I wish that we could now re-examine the material of the last three lectures or of some other portion of the history of science in order to display again the correspondence of this description with the facts of research. But perhaps you will apply this test for yourselves. Here I will simply express my own conviction that science has in fact progressed by a series of circular attempts to apply differing orientations or points of view to the natural world.

I state this here as a matter of fact, but I think it also a matter of necessity. I do not believe that the human mind can work in any other way. And in the next two lectures I will discuss certain reasons drawn from psychology and from logic for supposing that science could not advance at all without the benefit of such vague and qualitative predispositions to laws. But for the moment we need not debate the necessity of these principles of orientation.

Whether or not they are necessary, they normally exist. And the very fact of their existence is filled with consequences for the progress of all scientific research.

This leads me to some remarks on a subject which we may, by analogy, call the dynamics of scientific ideas, and this topic may be rephrased as the study of the manner in which scientific orientations, and the theories associated with them, are altered in time and ultimately replaced by new and radically different orientations.

We have just stated that scientific progress arises from a process in which points of view derived from older speculative sources, or from common sense, or from other parts of science are applied to a particular group of natural phenomena. This attempted application, if it is destined to have any significance in the history of science, enters almost immediately upon what has recently been called its classical period, a period in which it is eminently successful in unifying phenomena previously thought to be disparate and in suggesting new observations and new experiments whose results in turn fit the theory. During this period the original vague orientation is itself altered, refined, and made more precise. It is productive of laws, which may, though they need not, be quantitatively formulated.

In short, the classical period of a scientific orientation is characterized by the production from a vague and speculative point of view of a more or less precise scientific theory whose laws may be stated in scientific texts and verified by prescribed operation. And simultaneously the classical period is marked by the extension of these new theories to aspects of nature which had not originally been considered to fall within their jurisdiction. It is a period of triumph in precision and scope; and it is a period in which the orientation and the theories that have resulted from it become rigid and normative for the profession.

In our discussion of the phlogiston theory, we examined such a period in considerable detail. It was the period in which the alchemical element or principle, fire, was transformed to the subtle fluid phlogiston, and in which phlogiston became not merely a principle of combustibility but equally one of metallicity and a subtle effluvium which poisoned air. In this period the classificatory principles which separated phlogiston-containing substances like carbon and the metals from substances like air and the nitrates—which could absorb it—was fruitfully extended. They provided, you will remember, an immediate position for the gas hydrogen, as soon as it was rediscovered by Cavendish. And they led scientists like Scheele and Priestley to the discovery of oxygen and of photosynthesis in the search for dephlogisticating agents.

But scientific theories and scientific points of view never seem to remain indefinitely in the classical period. Gradually their fruitfulness is exhausted. All the natural phenomena which can be readily assimilated to the theory are embraced by it. And so by a transition which is usually impossible to mark precisely, the now rigidified professional orientation leads from triumph to previously unobserved difficulties. These difficulties may be of a number of sorts—they may arise from the more refined techniques of measurement whose application displays the existence of deviations from the law; they may arise from a series of genuinely new observations which were themselves suggested by the old orientation but which do not seem to fit it; or they most often arise from an attempt to embrace within the new theory a set of observations which had been known for a long time but which had been ignored during the period when there were easier directions in which to pursue research.[7]

These difficulties, these departures from the expected behavior of nature ought to destroy the theory on which the predictions were based. But, interestingly enough, they never do so. For scientists are always reluctant to abandon theories or points of view which have been fruitful in the past, and they are always provided with a number of alternatives to such a rejection. They may simply dismiss the apparently discordant observations and claim that a more carefully performed experiment would have eliminated the appearance of discord, or that the difficulty, though not understood, is trivial. They may accept the result of the experiment but claim that it does not in any way invalidate their theory, that on the contrary the failure is in another theory which was used in the construction of the experimental apparatus. For no single scientific theory can ever be tested without relying upon laws drawn from many other portions of science. Science is an interlocking fabric.[8] The determination of the position of a star by a contemporary astronomer involves not only a knowledge of the relevant portions of mathematics and astronomy, but also an understanding of the physical optics embodied in the construction of his lens system, the mechanical principles embodied in the construction of the mount of the telescope, the chemistry which determines the properties of his photographic plates, and the electricity which runs the motor which keeps his telescope pointed to a certain part of the heavens. A single stray observation or a group of consistent stray observations may be produced by or may be blamed upon a shortcoming in any one of these. It certainly need not reflect upon the astronomer, of which he himself will always feel the most certain. Frequently this stubbornness, this unwillingness to admit the existence of an error in one's own theory pays off, for the sources of the error do not always lie in the field of the experimenter who first discovers them. It was, for example, the astronomer Bradley who by a careful investigation of certain anomalies of astronomical information first discovered the aberration of light and convinced the scientific world that light possessed a finite velocity.

But even when further investigation of the possible sources of error fail to reveal any shortcomings in the equipment or in the portions of other scientific theories which went into constructing the equipment, the scientist still need not abandon his own theory. He can always adopt some small ad hoc modification of it. Any theory can be adapted to such new facts without really important distortions.

This was clearly illustrated in our study of the phlogiston theory. After the middle of the eighteenth century an increased knowledge of the properties of gases and an increased conviction that all chemical substances must have weight provided numerous problems for any believer in the theory. But the theory was not abandoned. Some chemists declared the problems irrelevant; other blamed the equipment—as in attributing the gain in weight to fire particles let in by the glass; others saved the appearances by introducing small modifications of the theory itself. And there were any number of such modifications which would do the trick.

This is typical of what we may now call the crisis stage in the progress of a scientific orientation. It is a stage during which all professional scientists continue to adhere to the point of view and to the theories derived from it; it is a period in which the bulk of research is directed by the old orientation. But it is equally a period in which the theory itself has become cumbersome due to the accretion of numerous ad hoc hypotheses designed to save it, and it is characterized by a lack of unanimity among practicing scientists as to which ad hoc hypothesis should be adopted to explain the results of particular experiments. Everyone believes the theory, but none is quite sure what it is.

Some such crisis stage seems inevitably to precede the overthrow of a scientific theory and of the orientation which has accompanied it. For the overthrow of the old theory is never accomplished directly by the difficulties which the theory has encountered, but rather by the suggestion of some alternate orientation which it is shown resolves the difficulties. And this new suggestion does not necessarily proceed from a new experimental discovery. On the contrary, it is more often suggested by a re-examination of an old experiment which has achieved a new significance because of the particular difficulty in which the old theory is laboring.

We saw this occur in the case of Lavoisier, who, approaching chemistry for the first time during a crisis period, started his experiments with the combustion of phosphorus and sulfur. It was the large decrease in the volume of the gas and the large increase in the weight of the solid product in this experiment that suggested to him that combustion might better be accounted for by the effect of the fixation of a portion of the air than by the loss of a subtle phlogistic principle. Yet these experiments had been performed before, and this interpretation of the role of air in combustion was an old one. And I would suggest that it is to a great extent the existence of the crisis period in the conceptual understanding of combustion processes which accounts for the difference in the role played by the experiment at the earlier and the later date and for the difference in the reception accorded the interpretation.

We have already remarked that scientific theories are not overthrown by internal difficulties but buy new scientific theories and new orientations. But here we should also remark that scientific theories are not overthrown at all during their classical period, that a scientific theory which to a modern critic appears the better or the correct theory is unlikely to be proposed or if proposed is unlikely to be accepted until the theory to which it is in opposition has itself run into internal difficulties. And this is true even when all the evidence on which the subsequent theory is based is already in existence.

But the mere suggestion of a new orientation toward the groups of phenomena formerly handled by the old theory does not automatically lead to the overthrow of the older set of conceptualizations. For the new orientation must first gain the precision and show the fruitfulness which characterized the older theory in its classic period. And meanwhile the theory and its author will meet continual opposition from the adherents of the older conceptual scheme, an opposition in which few of the techniques of personal vilification and social ostracism have not at one time or another been employed. Galileo may well be the only scientist who, until the last decade, has been imprisoned for pronouncing scientific opinions but he is certainly not the only one who has been vilified and who has lost professional prestige and the opportunity for professional advancement.

I find that it surprises most people to discover that scientists behave so much like other people when their pet theories are attacked. And this surprise is understandable because from the point of view of textbook science a conceptual revolution, the overthrow of an old theory and the establishment of a new one, always appears as a simple addition to the sum of human knowledge. The old theory as represented in the textbook enabled the scientist to predict and control a certain group of natural phenomena. The new theory as described in the new textbook enables the scientist to understand all of these phenomena and some new ones besides.[9] Before Lavoisier chemists did not know how to handle quantitatively the loss in volume of a gas and the gain in weight of the solid upon combustion. After the chemical revolution they could control everything which the phlogistic chemists had controlled and more besides. The revolution had simply added to the stock of scientific knowledge.

But from the point of view of a working scientist, a scientific revolution is always very nearly as destructive as it is constructive. As I have emphasized throughout these lectures, the older theory had proceeded from and developed with an orientation toward the operation of nature whose effect was to suggest the particular importance of certain aspects of the phenomenal world and to deny the relevance of others. Further this was a viewpoint which emphasized certain metaphorical abstract connections between phenomena which from another point of view appeared totally disparate. But the scientific revolution, while preserving the predictive content of the older textbook science, did not at all preserve the content of the older point of view toward nature. It demanded the destruction of old analogies in favor of a set which was new and strange. It may have suggested new sorts of entities underlying the phenomena previously studied. And it took as its point of departure a detailed study of an aspect of nature which for the older orientation, at least in its classical period, presented no problem at all. We may phrase this by referring to the completeness of a scientific orientation self-contained. The problems it suggests are solved within it. The new problems, from which the subsequent orientation will proceed, do not exist for it.[10]

Let me remind you once more. Lavoisier founded the chemical revolution upon the analysis of the weight and volumes of substances entering into and produced by chemical reactions. Since his time no one has doubted the relevance of these techniques to chemical analyses. Yet before his time chemistry had proceeded a long way; these experimental techniques had been available to it, had been applied at random; but had not usually been considered relevant to the determination of chemical theory. It was only by a long drawn struggle in which Lavoisier and his adherents convinced the chemical profession of the pragmatic value of considering weight relations to be fundamental rather than accidental characteristics of chemical reactions that the chemical revolution was completed.[11] But through the chemical revolution a group of experiments whose results had been conceded for centuries became suddenly points of departure of a new theory; an aspect of nature previously judged secondary became a primary tool of chemical analysis.

I suggest then that it is only in retrospect that a scientific revolution can be viewed as a net addition to the sum of human knowledge, and that from the point of view of the man involved in it, it is always about as destructive as constructive. For it demands the abandonment of the perspective in which he has so far viewed natural phenomena; and the adoption of an alternate perspective, which to the extent that he has accustomed himself to the old one, must always appear as a distortion of scientific vision. And this, I think, accounts for the battles which rage at the time of conceptual reorientation in science.[12] Even here of course one sees large variations in the degree of reluctance to accept new conceptualizations, but these variations in degree can themselves normally be explained by the variation in the magnitude of the conceptual transition required by the revolutionary shift. Dalton's application of atomism to chemistry simply adapted to chemistry a metaphysical mode of thought which had provided the foundation for a great deal of physical science for a century. Portions of atomism were implicit, and in Lavoisier's case explicit, in the chemist's description of what we should now call physical changes, changes of state. And so Dalton's proposal was almost immediately accepted by the chemical profession except, and the exception was again significant, that Dalton's work was in fact rejected by a group of French chemists surrounding Bertholet, who believed that the chemical elements can enter into chemical combinations in any proportions. For this belief they had found experimental evidence, evidence which we should now say they had misinterpreted, just as Dalton had actually misinterpreted the forces at work in producing homogeneous mixtures of the gases of the atmosphere. Again by taking different aspects of nature as fundamental to chemistry.[13] But Avagadro's hypothesis, which is now just as fundamental chemical thought as Dalton's, violated the fundamental conceptual tenets of all atomism in claiming that the atom was divisible. And this hypothesis had to wait forty years to be seriously considered. And that forty year period was one in which the difficulties of applying an atomic theory consistently in chemical analysis became so grave that the atomic hypothesis itself was very nearly abandoned. No wonder that at the end of the period the necessity of preserving the atom involiate seemed less a desideratum.

Now I believe that by focusing our attention on the destructive impact of a new scientific orientation, upon the points of view which lie beneath the preceding advances in science, we can understand somewhat more clearly certain of the recurrent characteristics of the history of scientific ideas. One of these was recently noted by Max Planck in his posthumous autobiography. He there observed that a new scientific truth does not normally receive recognition by proving to its adversaries either its validity or its value, but that on the contrary new conceptualizations enter science only when their adversaries have died, and a new scientific generation is enabled to make a fresh start.[14]

Charles Darwin put the same point particularly cogently at the conclusion of *On the Origin of Species*. He said: "Although I am fully convinced of the truth of the views given in this volume, I by no means expect to convince experienced naturalists whose minds are stocked with a multitude of facts all viewed, during a long course of years, from a point of view directly opposite to mine. A few naturalists," Darwin continues,

> endowed with much flexibility of mind, and who have already begun to doubt the immutability of species, may be influenced by this volume; but I look with confidence to the future,—to young and rising naturalists, who will be able to view both sides of the question with impartiality. Whoever is led to believe that species are mutable will do good service by conscientiously expressing his conviction; for only thus can the load of prejudice by which this subject is overwhelmed be removed.[15]

I find this statement particularly revealing because it points out three separate aspects of the problem presented by the existence of orientations. It suggests that people who have dealt repeatedly with the same facts from another viewpoint are particularly unlikely to be able to see the pragmatic value of relinquishing their viewpoint in favor of a new one. For the new viewpoint is one which sees different significances in the same facts and therefore in some sense makes them different facts.

But the statement goes further. It adds that mere open-mindedness or mental flexibility will not be sufficient to enable naturalists to see the value of Darwin's new thesis. They must already be convinced of the existence of difficulties in maintaining the thesis that species are fixed. They must be aware of the existence of what I have here called a crisis state. And, says Darwin, the most effective means of proselytizing for the new thesis is not to uphold the thesis itself, but rather to increase the intensity of the crisis by conscientiously expressing the opinion that species are mutable. If the crisis is sufficiently severe people will come around by themselves.

Finally, Darwin suggests that it is the youth of the profession, the group that have not accustomed themselves to seeing the facts from a particular viewpoint, who will be best equipped to weigh the relative merits of the two theories. Now the peculiar role of youth in scientific activity has been noted frequently in another and even more important connection. It is almost invariably the very young members of the scientific profession who make conceptually original contributions to science.

The evidence on this point, at least for physical science, is reasonably conclusive though not statistical. Galileo was about nineteen when he first noted the isochronous properties of the pendulum, and this observation, you will remember, was not simply one more fact but rather a new sort of fact around which the truly original portion of Galileo's mechanics was built. Newton was twenty-three and twenty-four during the famous years at Woolsthorpe; the years in which he initially formulated his views on the differential calculus, the theory of color, and the identity of the forces acting upon the apple and the moon.[16]

And in the twentieth century, when scientific education requires so much longer, the same phenomenon recurs. Bohr proposed his famous model of the atom at twenty-eight; Einstein was twenty-six in 1905, the year in which he advanced his theory of Brownian motion, the photoelectric effect, and the special theory of relativity. Heisenberg's matrix mechanics was published when the author was twenty-four; and Dirac's relativistic electron theory was published when he was twenty-six.

Of course there are exceptions: Planck was forty-three when he suggested the law for black body radiation. Schrödinger's wave equation was published when its author had reached years of scientific senility—he was thirty-eight. But Schrödinger's most important contribution was the mathematic formulation of a more speculative suggestion provided by his younger contemporary, de Broglie.

The pre-eminence of youth in scientific discovery has normally been accounted for by the greater flexibility, the greater openness, of the youthful mind. Yet I wonder whether this generalization is truly helpful. For the proverbial conservatism of age does not normally derive from a change of opinion on the part of the individual septuagenarian. On the contrary, the grandfather is likely to have held to the opinion of his youth at a time when his grandchildren have adopted more radical opinions. And it is my impression as a teacher that what is called the open mind of youth exists, if at all, only with respect to those areas of experience to which the young have not yet been exposed. Young men are at least as dogmatic as their elders with respect to those subjects about which they think they possess knowledge. And, like their elders, they suppose they have knowledge about those portions of experience with which they have to deal.

I wonder whether this may not be the key to the role of youth in science. The young man entering science has not had years of experience in viewing a selected range of facts in a given perspective. And this permits him, as he learns the facts, to choose between perspectives or on occasion to discover new perspectives of his own.[17] And this lack of commitment to any particular perspective is not, I think, at all the same thing as open-mindedness. For these students the facts themselves do not exist. As they are acquired they will be arranged in some perspective, and once they are so arranged the perspective will be difficult to shake.

I would contend then that emptiness rather than openness characterizes the youthful mind. And one indication of the validity of this view lies, I think, in noting that in those rather rare cases where fundamental new scientific insights are provided by older men, they are frequently provided by men who are approaching the field for the first time. Both Dalton and Lavoisier, for example, were past thirty-five when they advanced the views for which their names are known. Yet both seem to have gotten the inspiration which led to their revolutionary theories within two years of the time that they first engaged in chemical research. Until this time they had had little or no knowledge of chemistry.

But we need not stop here in our study of the effects of the dominations of scientists by orientations or points of view toward an area of scientific experience. Many other aspects of scientific progress reflect the same phenomenon. If for example I am right in suggesting that the crisis period in the development of a particular scientific conceptualization clothes old facts and old manipulations with new significances, then one would expect that these might equally be periods in which the same forward revolutionary step was taken by a number of scientists simultaneously. And indeed the occurrence of multiple simultaneous discoveries is too well recognized to require further comment here. Galileo and Descartes independently stated the law of uniform acceleration. Dalton's atomic theory was in many respects anticipated by Higgins. In our own time, Schrödinger's wave mechanics and Heisenberg's matrix mechanics were announced to the world in the same year.

Again I think we no longer need be quite so surprised to discover that periods of rapid advance in one science are not necessarily periods of rapid advance in another apparently closely related science. For we have already noted in examining the effects of the atomist cosmology that a point of view capable of guiding fruitful research in one field may at that time be highly unsuited to scientific advance in another. And this suggests that the so-called "unity of science" may not be an unequivocal blessing. Examples like these of the role played by orientation or viewpoint in the progress of a particular science can be multiplied almost *ad nauseum*. But perhaps the direction of these remarks is already clear. I should suggest that what we are now approaching is a clearer understanding of the causes and the nature of the phenomenon so often described by the phrase "a scientific discovery must fit the times" or "the times must be ripe."

Of course we have scarcely begun to explore either the nature or the sources of this phenomenon. Crisis periods in particular sciences proceed from a variety of sources many of which we have not touched upon this evening. They may, for example, be produced by social forces external to science. They may proceed from changes in economic structure which alter scientific motivation.[18] Or they may be produced by changes in political philosophy or in speculative cosmology. But these remarks if continued could only serve as an introduction to a full-scale study of the sociology of science and of the dependence of science upon the extrascientific climate of opinion. And I must here restrict myself to the considerations already advanced regarding the impact of professional orientation upon the professional scientist. These effects, positive and negative, are still very much with us.

This brings me (close) to the end of the hour and to the end of this evening's remarks. But before you go, I should like to supply you with an optional homework assignment. In the next lecture I want to study somewhat more closely the anatomy of orientation, and I will commence that study with a paradigm which will be more effective if seen in advance. May I introduce to you an amusing and illuminating puzzle?[19]

Notes

1. In Kuhn's script this sentence begins, "So that since . . . ." Another example of Kuhn's efforts to qualify and reduce the strength of these "prejudices" in this lecture is to frequently strike his use of the word "metaphysical" in earlier drafts. "Metaphysical orientation," for example, became just "orientation." Kuhn did not fully retreat from the word "prejudice," however, if only because he quotes Darwin, below, noting "the load of prejudice" his theory of evolution faced.

2. Kuhn annotated the text here: "Insert comment on Myerson as precursor of such study in final version."

3. Here Kuhn placed in brackets the sentence, "Are the various positions or points of space different in and of themselves or can they be differentiated only by the matter which occupies them?"

4. Kuhn annotated his text here, "Get in here later the Moliere doctor." Kuhn was likely thinking of Moliere's *The Imaginary Invalid* and its doctor who points to opium's "virtus dormitiva" to explain why it causes one to feel sleepy (Act III, scene 14).

5. Here Kuhn crossed out the qualification "a large portion of."

6. Here Kuhn wrote, "Add remarks on induction at this point later."

7. Here Kuhn noted "Logical difficulties, *vide* the difficulties in the Aristotelian categories."

8. A handwritten edit suggests that Kuhn may have intended "interlocked" instead of "interlocking." Additionally, the remainder of this paragraph is set off in brackets, indicating that Kuhn may have considered it optional, though it is not explicitly marked as such.

9. Kuhn placed the following two sentences in brackets, possibly for omission, and additionally replaced two appearances of the verb "understand" with "control."

10. Kuhn first typed "within it" but substituted by hand an illegible word which appears to be "for."

11. Kuhn placed this sentence in brackets.

12. Kuhn placed the remainder of this paragraph in brackets.

13. Here Kuhn typed "(the preceding out)," with the blank spaces perhaps indicating that he should elaborate or extend the somewhat telegraphic points made in these, and perhaps the following, sentences.

14. See Max Planck, *Scientific Autobiography and Other Papers*, trans. F. Gaynor (New York: Philosophical Library, 1949), 33–34.

15. See Charles Darwin, *On the Origin of Species*, ch. 15. I have corrected the abbreviated title (given as "The Origin of the Species") and the quotation where Kuhn originally typed "just opposite to mine" instead of "directly opposite."

16. Kuhn refers to Newton's family home to which he returned from Cambridge University during acute years of the plague. For a more nuanced account of Newton's so-called Miracle Years, see Richard Westfall, *Never at Rest: A Biography of Isaac Newton* (Cambridge: Cambridge University Press, 1983), 142–43.

17. Kuhn ended this sentence with a question mark, omitted here.

18. Here Kuhn added, "Attn. Sen. McCarthy," but crossed this out by hand in his typescript.

19. Noting that he should at this point "Ad lib from blackboard," Kuhn presented the so-called mutilated-chessboard problem, usually ascribed to the philosopher Max Black. In his book *Critical Thinking: An Introduction to Logic and Scientific Method* (New York: Prentice-Hall, 1946), 157, Black presented the problem among other "exercises in reasoning": "An ordinary chessboard has had two squares—one at each end of a diagonal—removed. There is on hand a supply of 31 dominoes, each of which is large enough to cover exactly two adjacent squares of the board. Is it possible to lay the dominoes on the mutilated chessboard in such a manner as to cover it completely?"

A chess or checkerboard with two squares of the same color removed from opposite corners.

Lecture VI

Coherence and Scientific Vision

At the end of the last lecture, I suggested that we might profitably approach the more detailed study of orientation by means of a mathematical puzzle, and I should now like to examine the puzzle I then posed for you. For this puzzle can serve quite successfully as a paradigm of many of the effects of orientation which we have already observed.

I have altered the figure slightly, or more precisely I have colored the alternate squares in the array so that the diagram suggests a checkerboard with the two corner squares now missing. The problem, you will recall, was to determine if it is possible to cover the checkerboard completely with 31 dominoes, each of which will cover any two adjacent squares.[1] Now regarding this diagram as a checkerboard instead of as a simple array of squares we can solve the problem directly. For a checkerboard, you know, consists of 64 squares, 32 of which are red and 32 black, and you note immediately from the diagram that the diagonally opposite squares are always of the same color. Therefore in our diagram in which the two corner squares have been omitted, we are left with a total of 32 red squares and 30 black squares, and we are asked to cover these with 31 dominoes, each of which will cover exactly two squares.

But now we see directly that there is no way of accomplishing this covering operation, for each domino must cover two differently colored squares, one red and one black. If we lay out 30 of our dominoes in any fashion whatsoever, we will cover just 30 black squares and 30 red squares, and we will be left with two red squares to be covered by the remaining domino. But there is absolutely no way of covering two red squares with one domino, so this problem possesses no solution, and we have proved this result.

There are many other proofs of this result. Most of them are more rigorous than this one and all of them are less simple-minded. They do not involve reference to checkerboards. But this one is, I think, the shortest and the most direct. It is the proof which most succinctly directs us to the heart of the problem. And this proof is one which we arrive at by taking a new point of view, a new orientation, toward the problem presented by the figure. This point of view is not itself suggested by the figure; there is nothing in the original diagram which indicates that we must look at two sorts of different squares. Yet until we look at the figure in this way we are at a loss for a method of attack.

And this is entirely typical of mathematical and physical discovery. The long proof, the logical proof, the rigorous proof, comes after the discovery; its effect is to put the content of the original discovery into a standard textbook form. But the original discovery is made by a route like that we've used in proving the checkerboard problem. It is achieved by applying a new point of view, a new angle of vision to an existing phenomenon and by gaining through this new orientation a totally new notion of the significance of the figure which has been there throughout.

Elementary as this paradigm of the checkerboard is, it shares a number of characteristics with the more elaborate sort of reorientation that we have examined in my earlier historical lectures.[2] It is in the first place a reorientation produced by attention to a new sort of problem. Until we reconsidered this problem there was no need to regard an array of squares as a checkerboard. Again the source of our new point of view lay in our own previous experience. That is, we gained our new insight into this problem by applying to it our pre-existing knowledge of the checkerboard, and this conceptual unification of two sorts of experience which we had previously seen as separate adds new significance to both. We see, because of this unification, new potentialities both in the array of squares and in the checkerboard.

Third, this new point of view toward the original array of squares opens up to us a whole new set of problems and solutions which were not in themselves suggested by the original problem taken alone. For having found this technique we need no longer restrict ourselves to two-colored boards or to rectangular dominoes. We can color our board with three colors or four colors; we can apply these in various orders; and we can cut our movable pieces or dominoes into figures that are three squares by one or that are L-shaped. In this matter we open up a whole new field of research. We have founded a new branch of mathematics. And this branch is one that did not exist until, under the impact of a particular new problem, we achieved a particular new orientation.

The history which we've covered has provided us with a number of examples of precisely this sort of behavior. Remember, for example, Dalton's discovery of the chemical law of multiple proportions. Before Dalton's time chemists had acquired a great deal of data which could have been used for the derivation of this law. But they did not see the laws exemplified by their data and this is not surprising. Examine for example the following figures.[3] There are two gases composed solely of carbon and oxygen. The first of these, carbon dioxide, has before Dalton's time been shown to be composed of twenty-eight percent carbon by weight and seventy-two percent oxygen. The second had also been analyzed and was found to consist of forty-four percent carbon and fifty-six percent oxygen. We can tabulate these figures and note in them no suggestion of simple whole number ratios.

composition by weight	carbon dioxide	second gas
carbon	28	44
oxygen	72	56

Now Dalton approached these figures with a new point of view, derived from a new problem. He was, you'll remember, concerned to explain the lack of stratification of the atmosphere, and he had decided that the uniform mixture of the gases of the atmosphere could be explained if these gases were supposed to consist of elementary particles of different sizes and different weights. But if, he said, the chemical elements consist of such different elementary particles, then chemical compounds must consist of these same elements united in simple whole number ratios. And this suggests that we examine the old figures in a new light. If forty-four percent of carbon unites with fifty-six percent of oxygen in the formation of this second gas, then how many parts of oxygen would unite with forty-four parts of carbon in the formation of the first gas?

This is a very simple algebraic problem. Forty-four twenty-eighths of seventy-two is just one hundred and thirteen. So that we can retabulate these figures as follows: forty-four parts of carbon unites with fifty-six parts of oxygen to form carbon monoxide; forty-four parts of carbon unites with one hundred thirteen parts of oxygen to form carbon dioxide. And one hundred thirteen is very nearly twice fifty-six, which gives us just the simple whole number ratio, the law of multiple proportions.

composition by parts	carbon dioxide	carbon monoxide
carbon	44	44
oxygen	(44/28)x72≈113	56

Here again we have the new problem created by the recognition that air is not a simple, a single gas. This new problem caused Dalton to apply to chemistry the old speculative atomistic point of view. The new point of view lends new significance to old data, and, I need scarcely remind you, this new point of view transformed all of chemical research. It is not restricted to its application in Dalton's original problem or to the discovery of the law of multiple proportions.

And, of course, the orientation serves one further function. For 113 is not quite twice 56. Twice 56 is 112. And it is in part the atomic orientation which provides a predisposition to the law that Dalton is able to judge that the disagreement is fortuitous, that these data prove the law.[4]

So far in this lecture I have suggested that the example of the checkerboard and the example of the law of multiple proportions provide us with paradigms of the role of orientation in scientific research. But there is another respect in which they are not typical, and it is to this that I should now like to turn. The examples which we've considered so far are perceptually of extreme simplicity. Both the array of squares and the numbers which we've examined on the blackboard were fixed perceptual entities. Changing our point of view may have led us to see new significances in the figures, but it did not alter the figures which we saw on the board.[5] At least this was the case to the extent that our change in orientation did not lead us to notice the deviation of the sides of the squares from mathematical straight lines or the slant of the symbols I employed for the digits in the tabulation. In these cases we were dealing with simple figures, and our perceptions were governed by an unequivocal code which directed attention to certain standard aspects of the symbols.

But when we deal with objects of the external world, with stones or metallic ores or plant specimens, we are dealing with entities of an infinite perceptual complexity, and there are no such unequivocal directives for our perceptions. In this more general case, there is an increasing body of psychological evidence that the point of view from which we approach these entities actually conditions what we see. That is, we now have reasons to believe that what we see when we look at an object is determined not only by the characteristics of that object but also by our expectations and our previous experience with objects of that sort.

You have all had experiences which will illustrate my meaning. All of you have at one time or another awakened from a dream to discover a strange, threatening figure crouching in the corner of the room. Yet another look convinced you that the threatening figure was really the familiar easy chair with a quilt piled on it. Then the figure lost its threat and you laughed at having been fooled. But were you really fooled? Did you really see the same thing on both occasions and did you simply misinterpret what you saw when you took it to be a threatening figure? Those of you who had such an experience recently may have difficulty in accepting the second, more usual explanation. For once you had discovered that the apparent threat was the usual chair with a quilt thrown on it, you also noticed a number of folds and humps in the quilt which ought to have prevented your ever seeing it as a threat. And you may wonder, remembering the intensity of your experience, whether you really saw those folds at all during the period when you were frightened. Didn't you really see different things in the two cases?

Apparently either of these interpretations may be applied to this psychologically extremely complex incident. You may either say that you saw the same thing in both cases, but that you interpreted what you saw differently, or you may say that you actually saw different things, that what you saw was truly dependent upon what you took the object to be. Yet I think that there are reasons for preferring the second statement. Its implications more nearly correspond to modern experimental findings. But before examining these recent experiments let us make perfectly certain that we understand the rather esoteric difference between the two positions.

I am now holding something in my hand.[6] If I ask you what you see, those of you who are not too far in the back of the room will probably say that you see an orange, and you're quite right. It is an orange. But if instead of exhibiting this orange I had held up a reasonable wax facsimile, you would have given me the same response—you would have said that you saw an orange—and in that case you would have been wrong. And this raises a very difficult problem. Was what you saw, was the visual image in your mind, the image of an orange or the image of an orange-colored wax ball?

There are two normal answers to this question. The first one is the classical answer, the one which almost everyone has believed since the seventeenth century. According to it, you did not see orange in either case. In fact, you never see an orange. What you saw, or what was given in your visual image, was a spherical shape with an orange color and a certain texture. These elements you examined, and because of your previous experience with oranges, you declared that what you were seeing was in fact an orange. In one case you were right, in the other case you were wrong.

But this is not the only possible interpretation. And according to its modern alternate we can more accurately describe what has occurred by saying that in both cases you really saw an orange, that your visual image was the visual image of an orange. For you have seen a great many oranges, you are quite prepared to see them, and accordingly there is a wide range of different visual stimuli which would have brought to your mind the image of an orange, or which at the very least would have caused you to act and react precisely as though you had seen a real orange.

And here the very real difference between the two views becomes apparent. The first opinion, the classical opinion, states that what appears to your mind is simply a conglomeration of qualities which is in some sense a literal reproduction of the stimulus provided by the object, and it implies that you retain the privilege of interpreting this conglomeration of qualities to be an orange or not. And the second interpretation says that for a wide variety of stimuli you will act exactly as though you had seen an orange. You have no alternative interpretation. You will see what you are accustomed to seeing, and you will behave as though all the interpretation took place before the image was formed in your mind. Your visual images are not unique responses to unique stimuli.

And now we can examine some evidence which supports the second interpretation. In a recent experiment a number of unprepared college students were exposed to a group of silhouettes cut from a neutrally colored cardboard. Some of these were silhouettes of objects normally colored red, a lobster claw and a tomato; others were of objects normally colored orange, a tangerine and a carrot; and some were of objects normally yellow, a banana and a lemon. The subjects were then asked to match the color of the silhouette to that shown by a color wheel, a device on which they could make any color appear at will by simple manipulation. And the result showed that under most experimental conditions the students would select on the color wheel a color which did not correspond to the actual color of the silhouette, but which was displaced from that color in the direction of the usual color of the object represented by the silhouette. The lobster claws were matched to redder colors and the bananas to yellower colors. Apparently the subjects' visual images were in part determined by their previous experience with the objects represented by the silhouette.

An even more striking result is provided by another experiment in which the subjects were exposed to selected groups of five playing cards. In each group from one to four of the cards were printed with the color reversed. For example, one such group contained a red five of spades and another a black four of hearts. Students were exposed to one card in the group at a time. In each case they were first shown the card for an extremely short interval of time; this interval was gradually increased; and on each exposure the subject was asked to identify the card exposed. As the exposure was gradually increased, a time interval was reached after which the subjects would consistently identify the cards correctly. And on the average the exposure required to identify correctly the first incongruous card exposed was fifteen times as great as the exposure required for the normally colored cards. As the subjects' experience in identifying abnormally colored, or incongruous cards increased, the time required for identification decreased, but the difficulty in seeing correctly an unexpected or unfamiliar object was marked indeed.[7]

And it was a case of "seeing incorrectly." For from our present viewpoint the significant aspect of this experiment is not that it took so much longer to identify correctly the unexpected than the expected cards, but that, prior to the correct identification of the abnormal cards the subject reported that they had seen a normal card. In most cases, for example, the red five of spades would be identified on short exposure either as a black five of spades or as a red five of hearts. The subjects did not claim an inability to identify the source of the stimulus. On the contrary, for all but the briefest exposures, they knew what they had seen. And what they had seen remained for a long time what they might have expected to see. They saw normal playing cards even when exposed to abnormal playing cards. Their perceptions were altered by their expectations.

There are many other experiments of this sort, experiments which show how our perceptions of size, of color, or of distance are influenced by the sort of objects which we suppose that we are examining. One such experiment shows that we will judge the distance of a rectangular white spot to be less if we are told that it represents a calling card than we will if we are told that it represents a package of cigarettes. And other experiments show that our perceptions, what we see, depends not only upon our habits, our previous experience, and our expectations, but also on our emotional attitude toward the viewed object. We have more difficulty seeing the unpleasant and the disagreeable than the beautiful. And for this the classic case is, of course, our difficulty in seeing anything but the beautiful physical characteristics of the one we love.

Experiments like these have led a number of psychologists and philosophers to point out that the world which we see is actually a great deal less complex in its structure and organization than the totality of the stimuli which produce our perceptions. We see certain objects and relationships easily, and we see these for any one of a number of widely different stimuli. To other sorts of stimuli we do not respond at all. And this is equivalent to pointing out that the world of our perceptions is not uniquely determined by sensory stimuli but is a joint product of the external stimulation and of an activity which we perform in organizing them.

This point of view is frequently expressed by the statement that we all live in a behavioral world, which our own activities play a large role in creating. And this amounts to saying that the world of our perception is a world we have simplified in accordance with the needs of our behavior. It is entirely necessary that we do this, for a world in which every different stimulus yielded a different perception would be a world without any stable objects. In a phrase which William James applied to the world of the infant, the world in which every different perception would be just a "bloomin, buzzin confusion."

In such a world we would be incapable of acting. We would be able to discriminate between the orange and the wax orange and thus we would never be fooled. But also we would be forced to discriminate between every real orange, for real oranges are also different. And if we provided a unique response for the wax apple and another for the real orange, so we would provide different responses for each of the real oranges because they too provide different stimuli. But in such a world every perception would be unique, for no two stimuli from the same or from different objects are ever quite identical. And if every perception were unique and every response took account of the uniqueness of the perception to which it was a response, then we would be incapable of responding at all. For we would be unable to apply our awareness of the pragmatic content of one perception to another different perception. In short, in a world in which every perception is unique, we could not learn from experience.

So if we are to act, if we are to respond, if we are to deal with the problems presented by our environment, we must order our perceptions of that environment. We must cut the perceptual world up into a number of categories to which we can produce uniform responses. We *must* build a behavioral world out of the "bloomin, buzzin confusion" by discriminating among the infinite variety of stimuli which are presented to us just those groups to which we may safely provide uniform responses. And equally we must suppress sensory discrimination of differences which do not call for differing responses. We must learn to see oranges and lemons as distinct, and we may have to learn to see oranges and tangerines as distinct, but we must not discriminate between different oranges even though they supply different stimuli. And this, says modern experimental psychology, is what we do. We cut up the world of our perceptions to fit our needs.

Of course I am exaggerating when I say that we must not see the difference between oranges. Obviously we both see such differences and we respond to them. We can tell a squashed orange from an injured one, a large one from a small one, a navel orange from a normal orange. And we may react differently according to these discriminations. But unless we have been trained to sort oranges according to their ripeness, their size, and their goodness, we are relatively insensitive to their individual differences. A trained fruit rancher who has to deal with oranges will see all sorts of distinctions which we will fail to see.

Perhaps this seems dubious as regards oranges, but let me suggest an example with which you are all familiar. It is commonly said that to a Caucasian most Chinese look alike, and to a Caucasian it is quite natural that they should do so. We meet very few Chinese and the overwhelming majority of these are either restaurateurs or laundrymen. An ability to discriminate between them would be a useless perception and useless perceptions are confusing. On the other hand we must learn to discriminate the physiognomies of other Caucasians, for our entire social and economic life may depend on our ability to identify individuals and react accordingly. It may, though, restore our sense of proportion to remember that for the Chinese the situation is just reversed. He has difficulty, unless he has lived in the occident, in distinguishing many Caucasians.

So I believe, that we discriminate only those aspects of experience which are behaviorally relevant and that our behavioral world consists of these and these alone. Of course in particular instances we are capable of discriminations which are not relevant to our behavior in these instances. Certainly, for example, we can tell a red coat from a green coat although both are effective for keeping us warm, and warmth was our presumptive object in purchasing them. But there are many areas of experience in which the ability to discriminate between red and green is important. Telling red apples from green apples may save us acute discomfort. Telling red lights from green lights may save us a great deal of trouble in traffic court. And in societies where such color distinctions as these are not of importance, the ability to make such discriminations is normally lost. In the life of the Chookchee eskimoes, for example, color discriminations are of very small importance.[8] Accordingly the Chookchee, who have the same physiological perceptual apparatus as ourselves, have very poor color vocabulary and are unable to sort strands of wool which we find to be strikingly different in color. Yet these people do not lack visual acuity in the matters which concern them. They are reindeer-riding people, and they apply twenty-four names to the various patterns which they discriminate in reindeer hides. And to a visiting anthropologist most of these patterns are completely indistinguishable.

These examples suggest another important characteristic of behavioral worlds. They are not unique. The behavioral world of the Eskimo is poorer in color but richer in patterns than our own. Behavioral worlds are therefore in part determined by the social needs and conditions of the society for which they serve, and differing needs may produce differing behavioral worlds. But even within a given society there may be considerable difference in the behavioral worlds of different individuals and different professions. Remember the orange-sorter who responds to differences that are invisible to the normal member of his society. Or consider the naval lookout who is trained to see certain sorts of visual stimuli on the horizon. These are stimuli which exist for the untrained observer as well but which are not, by him, translated into visual images. There are objects in the lookout's behavioral world that do not exist for his lay contemporaries.

The same is true for every specialist—he sees objects and distinctions which are not available to the man who does not need them. And this suggests that behavioral worlds may never be criticized for falsity or inaccuracy. They are always poorer than the totality of stimuli which they order. They are never "true." They cannot be judged for accuracy, but only for adequacy. Do they or do they not contain sufficient ground of discrimination to support well-directed activity? Do they or do they not fulfill individual and social needs? Are they adequate to life?

Of course behavioral worlds are not always adequate, and when they are not we must learn to alter them, make a new sort of discrimination. And when we do this, we change our behavioral world. We alter what we see. In a famous experiment performed at the Hanover Institute, a subject is seated in a badly distorted room, a room in which the walls are not vertical nor the corners square.[9] And he is provided with a pair of specially designed glasses which correct for these distortions, so that the subject, wearing them, reports that he is in a normal room, that the walls are vertical and the corners square. The subject is then given a pointer and asked to touch first the upper right hand corner of the room and then the upper left hand corner. He has no trouble with the upper right hand corner, but because of the unseen distortion of the room he keeps missing the upper left. After repeated trials he does learn to touch the upper right and upper left corners successively. But when he has learned to do this he can no longer see the room as undistorted. In adjusting to the new environment he has been forced to change what he sees.

Another experiment performed at Dartmouth provides an even more striking indication of the manner in which we may alter our perceptions in order to make them conform to the needs of our behavior. In this experiment subjects are provided with a pair of glasses or goggles whose lenses completely invert the visual image.[10] The subject wearing such goggles initially finds himself seeing everything upside down. But a perception of this sort makes action difficult or impossible. A subject reaching for an object at the top of his visual field will always miss the object, for without the goggles the object would have appeared at the bottom. Successful purposive activity requires a readjustment to the new perceptual world presented through the goggles.

And this readjustment can be achieved. But when it is achieved the subject sees everything right side up again. Once he has made the adjustment required for activity with the goggles his perceptions are exactly the same as they were without the goggles. And this shows once again that many of the characteristics of our perception are determined by the necessities of behavior rather than by the nature of the stimulus or by the structure of our perceptual apparatus.

I have expressed the outcome of all of these experiments by saying that we see the same thing or respond in the same way to a variety of different stimuli. But these last two experiments make it obvious that I might equally well have said that we can respond in several different ways, or see a variety of different things, when confronted with a single stimulus. And this double aspect of the problem makes clear the great difficulty posed by these experiments in terms of a vocabulary which treats the individual perceiver as a camera combined with a responder. Our role is much more active than that, and the world of stimuli is far more plastic than we normally suppose. We make continual and profitable use of this plasticity in creating a behavioral world within which we can and do act.

I have dealt at very great length with the perceptual or visual aspects of the problem presented by the existence of a behavioral world because I believe that the experimental investigations in this area are, as far as they go, typical, unequivocal, and thoroughly surprising. They illustrate most clearly the inadequacy of our commonsense notion of our passivity as perceivers[11] of objective data about the external world. But there are other sources which display equally interesting and rather different effects of our participation in the construction of behavioral worlds.

One particularly interesting source of evidence on this point derives from the modern comparative study of language. During the last half century increasing experience in the analysis of non-Indo-European languages has made it increasingly apparent that differences between languages cannot be represented as simple differences in the particular signs or symbols which are applied in different languages to identical situations. Differences between languages may equally derive from different world perspectives: different languages may refer to different behavioral worlds. So there may be real contrasts between the objects and between the connections between objects in the different behavioral worlds displayed by various languages.

In different languages there are, for example, many differences between the perceptual complexes which are seen as objects or things and denoted by nouns, and the perceptual complexes which are seen as activities and are denoted by verbs. In English we distinguish sharply between nouns like "house" or "man" and verbs like "hit" or "run." Apparently this distinction is based upon a difference in the temporal duration of the perceptual complex denoted. Man or house endures; the activities, hit or run, are of short temporal duration. But this distinction is impossible to make precise. A man is continually changing. Running can endure for a considerable period. So it is not surprising to discover that in other languages the division is accomplished quite differently. In the Hopi language, for example, perceptual complexes of short duration, which we would denote by nouns, as, for example, "lightning" or "wave" or "flame" are all verbs in Hopi. Where we employ lightning as a noun, they say "lightning occurs" but with one word. And in the language of the Nootka Indians of Vancouver there are no such things as nouns. All these perceptual complexes become verbs. What we denote by "house" they denote by "it houses" or "a house occurs."[12]

Linguistic differences of this sort apparently reflect differences in the behavioral world of different societies. More precisely, they reflect differences in what we have previously called the metaphorical connection between different perceptual complexes available to different social groups. Situations represented as closely parallel by one language are seen as totally different when viewed through another. We are willing to compare perceptual complexes denoted by nouns or perceptual complexes denoted by verbs, but we will consider a request for a cross-comparison to be an absurd one. Yet experience with other languages shows that our judgment of the absurdity of such a request, our judgment of the difference between the two perceptual complexes, is governed by our language, by the way we cut up our world, and not by anything given objectively in the perception.

These differences in linguistic metaphor extend to far more complex situations. The late B.L. Whorf pointed out for example that the two situations which we should represent by the phrase "I pull the branch aside" and "I have an extra toe on my foot" are represented by almost identical phrases in Shawnee.[13] We have the greatest difficulty discovering any similarity between the two situations. But the Shawnee emphasizes the common form of forkedness in the two situations and the common excess over normal situations. And accordingly the two sentences which are formulated differently in the last four of the seventeen symbols required to represent them phonetically.

The Shawnee see a similarity in logical categories where we see none at all.[14] And of course these metaphorical connections implicit in our behavioral worlds are of tremendous importance in the search for regularities upon which we can predicate our behavior. Remember, for example, the transformation in the study of motion which was correlated with the change in the meaning of the word *motus* from a word denoting a change of state to a word denoting a state. Or consider any of the numerous other examples of metaphor which we discussed in the last lecture.

Let me supply you with one other example of the differences between behavioral worlds. It is drawn from the field of child psychology and therefore presents us with a rather different problem than the one implicit in our previous illustrations. For the words with which the children indicate how they cut up the world are adult worlds, and children's syntax is very nearly adult syntax, so it is easy to suppose that children use these words in the same way that we do and that they understand by them the same things as we do. That is, we are prone to assume that children are just little men and little women, and that their behavioral world is like our own.

But you all know that the maturity implied by the child's choice of words and by his syntax is at least very frequently completely misleading. In many areas of thought and of perception children operate quite differently from their elders, even though they verbalize the end-product of their thought and their activity with an adult vocabulary. You have all had illustrations of this, for you have all at one time or another found yourself operating at cross purposes with the child's mind. You have listened to a child's explanation of a given event or behavior and found the explanation meaningless—although to the child it was apparently totally satisfactory. Or, you have attempted to supply an adult answer to a child's question and have discovered that not only was your answer totally unsatisfactory to the child, but that the answer satisfactory to him was one you could not have discovered by any literal interpretation of the words which he used in asking for it.

In those areas of experience to which children apply thought they do not necessarily think the way we think. And though to us the mental connections which the child employs may appear nonsensical or the products of ignorance, they are for the child an adequate rationalization of his world. But his world, that is his behavioral world, is different from our own. As the child grows up these modes of thought are no longer adequate to deal with the developing complexity of the problems which he meets. The thought process and the concomitant behavioral world are changed, and the behavioral world at which he arrives is a joint product of the older society in which he is raised and of the new problems which he, as a member of that society, may have to face. It thus need not and usually will not be quite the same as the behavioral world of the preceding generation.[15]

I will not here attempt a reconstruction of the behavioral world of the child, but I should like to supply you with one example of an experiment in which children use an adult concept but with a different meaning, and of the manner in which the child's concept is proved inadequate and is changed. This illustration is drawn from the work of the famous Swiss child psychologist Jean Piaget, who has done more than any other contemporary investigator to illuminate the nature of children's mental processes. This example is particularly useful because of the similarity it shows to mental processes which we have already examined in our investigations of the history of science. Also, it raises a problem about the role of logic in science to which I will return next time.

Children use terms like "to go as far as" or "one object travels as far as another object." Under many circumstances the child will apply such adult phrases in the same way as the adult does. That is, he will apply them in the same cases as the adult does. But it can be shown that he does not normally grasp the same aspect of experience in making his judgment of the applicability of the term.[16] Consider, for example, two pieces of wire in the following shapes, each of which has a movable bead on it.

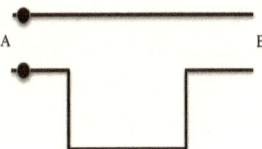

As the beads are moved from A to B along these two paths, the child will usually say that the one goes as far as the other, from A to B, without realizing the incongruity presented by the vertical portion of the second wire.

That the bead moves on the vertical sections, but does so without nearing its destination, is a problem that when pointed out will eventually lead the child to refine the concept of "go as far as" by discriminating between the total displacement as the beads move from A to B and the unequal distances each travels along its wire.

We could dwell at length on the parallels between this example and the medieval and ancient understanding of motion which we discussed in the second lecture. Here again we see motion grasped in terms of its endpoints; we see the logical difficulty presented by the attempt to analyze motion in these terms; and we have observed the way in which the modern grasp of motion is evolved through successive attempts to grapple with the logical problem. The parallelism between Aristotelian and children's physics is here and in other areas quite precise, but I find it impossible to be certain what significance we may reasonably attribute to it. Here I would simply point out that the parallelism exists and that it illustrates once again the temporary behavioral adequacy of an organization of perceptions other than the one which we customarily employ.

But though I think that we must equivocate about the significance of the detailed parallelism between the child's world and the Aristotelian world, I see no need for a similar equivocation as to the parallelism between what I have here called the behavioral worlds and the scientific orientations or points of view which we discussed in the last lecture. In spite of the vast gap between the sorts of scientific behavior which we have examined in previous lectures and the very simple experimental behaviors which we have discussed this evening, the similarities are too striking to be without significance. And I should like in concluding to provide you with my preliminary evaluation of certain of these significances.

We can now say, I believe, that the scientist, too, operates in a behavioral world. It is not the everyday or the common social behavioral world which we have examined this evening. It is rather one of a number of professional behavioral worlds which, in the more advanced sciences, legislates for an area of experience foreign to most laymen. But it nevertheless parallels the behavioral worlds we have examined in many important respects.

In the first place all these behavioral worlds are incomplete. The correspondence between the perceptions available in a given behavioral world and the mental categories in terms of which that world is treated is so close that all questions presented by the behavioral world can be answered in terms of the corresponding mental categories. In the last lecture we phrased this for science by stating that all questions arising within a given orientation or point of view had solutions within that point of view. And we have seen the same characteristic illustrated for behavioral worlds. The subjects who did not identify correctly the incongruous cards in the experiment discussed earlier this evening did not plead ignorance. On the contrary, they fitted that card into a preferred category in their behavioral world. The Eskimos who cold not sort strands of wool which to us are clearly different could not see any difference between the strands. They did not have the vocabulary or the mental categories requisite to the task. Their perceptual or behavioral world did not contain the diversity which could necessitate the task. The child who uses words differently than we, and who therefore finds no satisfaction in our answers to his questions, can nevertheless be given an answer which will be to him entirely satisfying. Questions for which no such satisfactory answers exist are questions he could not ask.

In the second place, behavioral worlds like scientific orientations can finally prove inadequate and require change. They too may be forced into crisis stages if they prove inadequate to mediate experience or if they prove to be possessed of logical inconsistencies. When the exposure to the normal cards was too long, the incongruity became apparent to the subjects of the experiment. They had to create a new classificatory category for the cards. But before they could achieve this new categorization, they had to go through a definite crisis period, a period in which they were aware that something was wrong with their earlier classifications but were unable to see quite what it could be. The man readjusting to the distorted room in the Hanover Institute experiment goes through a similar intermediate stage, a stage in which having recognized the inadequacy of his first perceptions he was unable to describe any that would replace it. He was forced to resort to almost pure trial and error to complete his task. And the child confronted with logical difficulties in reconciling his two notions of the meaning of the phrase "as far as" goes through a similar crisis. He first tries to preserve both meanings but to apply one and then the other, or he adopts small modifications not quite consistent with either in an effort to reconcile the two. And this parallelism between the psychological structure of crisis states in science and in behavioral worlds can be developed in considerable detail.

And finally, the transformation in a behavioral world produced by a crisis, by the recognition of an inadequacy in the older world, transforms experience as well as the mental category in terms of which we deal with experience. One can learn to recognize incongruous cards. After that, the category is as readily available as the category for normal cards, and the cards are immediately perceived. The man in the distorted room was able to reorganize his perceptions. But when he had, he saw a different sort of room. And the child who discovered the new use of the phrase "as far as" had to shift his grasp upon the phenomenon in order to do so. The same question now directs his attention to a different aspect of the experience.

In all of these cases, as in the scientific examples we have examined, the effect of the shift in behavioral worlds was destructive as well as constructive. The new behavioral world was undoubtedly an improvement. It was more adequate to experience, but it required the destruction of an old and a complete way of grasping experience prior to the reconstruction of a new behavioral world.

It is because of parallels like this, parallels susceptible of a far more detailed development, that I suggest we equate the notion of scientific orientation with that of a behavioral world. And it is in part because of the psychological necessity of some behavioral world as a mediator and organizer of the totality of perceptual stimuli that we will never be able to eliminate from the scientific process the orientations which originate in experience but subsequently transcend it and legislate for it.[17]

Notes

1. Instead of this sentence stating the problem, Kuhn noted only "Restate problem."

2. I have adjusted Kuhn's original phrasing, which was "the past historical lectures."

3. Here Kuhn instructed himself, "Put these on the blackboard." I have supplied the following tables and noted that (44/28)72 is not exactly equal to 113.

4. This paragraph replaced a longer discussion, the subject of which which, according to the parenthetical note that concludes it, was moved from this lecture to lecture V. This longer discussion reads:

But this physical example of a reorientation is different in one important respect from the geometrical problem which we examined earlier. And this difference suggests one other important function of a point of view. You will already have noticed that even after our manipulation of the data the figures arrived at do not exactly confirm the law of multiple proportions. Two times fifty-six is one hundred twelve, not one hundred thirteen, and in judging the law it is necessary to decide whether this difference marks a true deviation from the law or whether it simply notes a minor error due to some failure of the experiment or the experimenter. And this decision will normally be made upon the basis of the point of view which prevailed when the experiment was undertaken. Dalton expects the law of multiple proportions to hold; he examines the data and is delighted to find that they prove the law, although frequently the observed deviation from simple whole numbers is as great as twenty parts in a hundred, far greater than in our example.

Yet when Gay-Lussac proposed the law that gases united in ratios which provide simple whole-number volume relations Dalton rejected the law as fortuitous and pointed to the deviation of the experimental values from simple whole-number ratios to prove the validity of his rejection. Yet these deviations were in almost all cases very much smaller than the ones in the data which Dalton had used in proving the law of multiple proportions.

Here we see the point of view acting not as a source of new insight and new connection but as a mental or psychological framework which actually legislates for the external world. It is our orientation which enables us to say this sense of experimental observation shows a true law of nature, although in fact the experiment will always show deviation from what we take to be the law. And it is this same mental framework which leads us to reject other experimental evidence which though itself inconclusive would lead to a conclusion opposed to our law. And though these rejections viewed retrospectively frequently seem like sheer perverse conservatism as blocks to the progress of science, they also have an extremely constructive function. As we have noted again and again in the historical portion of this series, there are always facts which will contradict any proposed theory, and if one is to develop a new scientific orientation, if one is to explore and exploit a new scientific theory, one must simultaneously seek for aspects of experience which confirm the theory and find ways of explaining away those aspects of experience which seem counter to the theory. (Try to get most of this very last material into lecture number 5 and out of this one.)

5. Kuhn continued, but then crossed out, "or at [l]east this was apparently the case." Kuhn's ambivalence here about whether scientific change is driven by changing ideas or changes in the objective world itself (indeed, whether "objective world" is a coherent notion) anticipates similar ambivalence in *Structure* that would invite intense criticism of Kuhn's ideas, continuing in some quarters to this day. See, for example, Errol Morris, *The Ashtray (or the Man Who Denied Reality)* (Chicago: University of Chicago Press, 2018).

6. Kuhn notes here, "At this point hold up an orange."

7. Here Kuhn reports on the research paper "On the Perception of Incongruity: A Paradigm," by Jerome S. Bruner and Leo Postman, *Journal of Personality* 18 (1949): 206–23, which he later discussed in *Structure*'s chapter 6.

8. Kuhn offers no references or sources for this and ensuing remarks about anthropological and linguistic research on Eskimo color perception. Later in the lecture, however, he refers to and paraphrases linguistic observations of Benjamin Lee Whorf. Along with his mentor Edward Sapir, Whorf theorized about interactions between language and perception in ways that Kuhn drew on here and later in *Structure*. See John. B. Carroll, ed., *Language, Thought, and Reality: Selected Writings of Benjamin Lee Whorf* (Cambridge, MA: MIT Press and New York: John Wiley & Sons, 1956). For a critical overview of the Sapir-Whorf thesis and Eskimo color vision, see Laura Martin, "'Eskimo Words for Snow': A Case Study in the Genesis and Decay of an Anthropological Example," *American Anthropologist* 88(2) (1986): 418–23.

9. This and similar experiments were devised and performed by Adelbert Ames at Dartmouth College. See Roy R. Behrens, "Ames Demonstrations in Perception" in E. Bruce Goldstein, ed., *Encyclopedia of Perception Vol. 1* (Sage Publications, 2010), 41–44.

10. See George M. Stratton, "Vision without Inversion of the Retinal Image," *Psychological Review* 4 (1897): 341–60, 463–81. Kuhn cited Stratton explicitly in *Structure*, chapter 10.

11. Kuhn used the word "perceptors," instead of "perceivers."

12. Here Kuhn paraphrases Whorf's observations in his essay from 1940 "Science and Linguistics," reprinted in *Language, Thought, and Reality*, op. cit., 207–19, esp. 215.

13. See Whorf's essay of 1941 "Languages and Logic," in *Language, Thought, and Reality*, op. cit., 233–45, esp. 234.

14. Kuhn bracketed this sentence and the previous paragraph, possibly for omission.

15. Kuhn bracketed this and the two preceding sentences, possibly for omission.

16. I have constructed this figure and the remaining three sentences in this paragraph on the basis of the text and Kuhn's two notes to himself: "Draw a straight line and a lower square wave on board" and "On diagram, discuss fact that primary meaning of term is total displacement rather than distance traveled through. Then point out that this is not simply a transposition of definition. The child also recognizes the activity that necessarily takes place in moving anywhere. That he recognizes incongruity in the particular situation in which the bead is displaced only vertically in the square wave pattern, and that the recognition ultimately leads him to use the term in the adult manner."

17. In this sentence I inserted "because of" and the article before "orientations." I removed an additional "I suggest" for added clarity.

LECTURE VII

The Role of Formalism

In the preceding lectures we have talked at considerable length about the nature of the physical sciences without ever discussing those characteristics which to many people seem their most unique and essential attributes. I refer of course to their use of deductive techniques, to their logical and their mathematical structure. These two, logic and mathematics, may for our present purposes be grouped together, for, as many of you realize, recent investigations on the foundations of mathematics have shown that most or all of the theorems of mathematics can be derived from the more general principles of modern logic. It is to a preliminary study of certain of the roles played by logic and mathematics in the physical sciences that I should now like to turn.

The importance of mathematics and logic to the physical sciences has been recognized since the beginning of recorded history. Yet even our present limited and controversial understanding of the nature of the relationship between them is dependent upon a series of developments within mathematics and logic which have taken place during the past hundred years. The effect of these more recent studies has been to expand and to generalize the subject matter of logical and mathematical studies, and one supreme result of this generalization has been the emancipation of logic and mathematics from the study of the real or the apparently real world.

This emancipation has had incalculable consequences for science and for philosophy. The intellectual revolution which has accompanied it has no parallel in the period since the seventeenth century. And, fortunately for us, many of the essentials of this revolution can be grasped through the study of elementary examples. The brilliantly elaborated, highly technical, immensely abstruse structure of modern logic and mathematics were necessary prerequisites for the modern understanding of the nature of these disciplines. But, this insight once provided, many of the lessons learned from the efforts of modern logicians can be retrieved in older, more elementary examples. This recovery constitutes our program for this evening.

One important caveat is however required. Modern logic and mathematics constitute an imposing edifice. There is little or no controversy about its significance or about its permanence. But we are not going to examine this edifice itself; we are rather to inquire about its foundations. And in this inquiry our analogy to the construction trades proves ill founded, for the unanimity about the superstructure does not in this case imply a unanimity about the foundation. The edifice of modern logic is unquestionably more secure than its foundation. And the foundation remains the subject of controversy. So you ought not be misled by the elementary triviality of my examples to the supposition that the conclusions are equally elementary; you ought not suppose that the stability of the superstructure guarantees the unique stability of the particular set of conclusions about its foundation to be discussed this evening.

You are all familiar with the use of logical techniques in at least two contexts. The first of these is, of course, the classical syllogistic reasoning, exemplified by the famous syllogistic form Barbara. If we take as premises "All men are mortal" and "Socrates is a man," then it follows from the rules of syllogism that "Socrates is moral."[1] This is not the only syllogistic form. But it is typical, so we need not examine the others.

An apparently different case is provided by the use of logic in Euclidean plane geometry. It is difficult to see the geometry we learned in high school as purely syllogistic. But here again we are confronted with a number of axioms and postulates or premises from which by logical manipulation a large number of conclusions about triangles and rectangles and curvilinear figures can be drawn. As in the syllogism, we have premises and conclusions. One can, for example, deduce from Euclid's postulates and axioms about the properties of points and lines the Pythagorean consequence that the square of the hypotenuse of a right triangle is equal to the sum of the squares of the other two sides: $c^2 = a^2 + b^2$.

Neither of these examples has escaped modern criticism. Syllogistic reasoning, we should now say, is accurate as far as it goes, but it is far too limited and restricted. The syllogism is valid, but in many cases it is cumbersome. We no longer insist upon the desirability or possibility of putting all logical arguments in syllogistic form. We have discovered more suitable and more general forms for logical derivations. And with the development of these more general logical categories came an attempt to develop mathematical subjects, like Euclidean geometry, in more rigorously logical terms. It had always been recognized that the connection between Euclidean axioms and Euclidean theorems was in some sense logical. But the only rigorous logic known was the syllogism, and the connection could not be made by the syllogism alone.

It turned out that Euclidean geometry, as developed in antiquity, or as we learned it in high school, is by no means the model of logical rigor that we have taken it to be. On closer examination, the axioms and postulates of Euclid prove to have little logical significance. Some are redundant, others meaningless or useless, and taken as a group they are entirely inadequate to the development of the theorems formerly deduced from them. In order that they might provide an adequate basis for the logical derivation of the theorems of Euclidean geometry, the axioms and postulates provided by Euclid had to be restated, and they had, in addition, to be supplemented by a large number of additional axioms which had previously been implicit in the procedures used in proving the theorems.

But this revision proved possible. New and complete postulate systems for Euclidean geometry and for other portions of mathematics were found. And this revision, though it clarified much that had been obscure in the original development of geometry and though it facilitated many new discoveries and generalizations, did not alter the fundamental form in which geometry had previously been cast. The axioms were different, both in their form and content; the logic was more general; and many of the theorems were new. But the geometric proof still proceeded from axioms by logic to theorem and theorems were still concluded with the old q.e.d. So in my further remarks about geometry it should be understood that I am referring to it in one of the twentieth-century revisions which make it complete.

Crude as our present examples are, they illustrate quite adequately two extremely important characteristic of all logical or mathematical systems. The first of these is the apparent necessity of the connection between the premises and the conclusions. If you believe that "Socrates is a man" and that "All men are mortal," then you cannot avoid the belief that "Socrates is moral."

Similarly if you accept the revised postulates of Euclidean geometry you must accept those theorems like the Pythagorean theorem which follow from the axioms. You may reject the conclusions; it may be wrong. But then you must reject the postulate as well.

The apparent necessity of the connection between premises and conclusions in logical and mathematical arguments is closely related to a second characteristic of these systems. The validity of the conclusions from the premises is, unlike the validity of a judgment about a matter of fact, entirely independent of any knowledge we may have about the real world. It does not depend upon anything we have learned through our senses, so there can be no errors of interpretation. And this aspect of the problem is frequently expressed by saying that mathematical and logical validity depend only upon the form and not the content of the statements which are employed in these disciplines. All that is logical in a logical proof can be abstracted from the content of the statements involved in the proof.

The power of the syllogism derives from exactly this characteristic. The truth of the syllogism considered as a whole is independent of anything we know about Socrates or of what we mean by mortal. The nature of the syllogism is equally well and more generally illustrated by writing "All A are B," "C is A," therefore "C is B."[2] The syllogism is valid or true whatever we take A, B, and C to be. If you want to know whether it is true that Socrates is mortal, then you must discover whether the premises are true. Then you must know something about men and about Socrates. But the syllogism itself is independent of any such information. It achieves its generality and apparently its necessity by being about nothing at all. It supplies us with no information.

Now mathematics turns out to have exactly this characteristic. In the Euclidean geometry which we learned in school, we spoke of points, lines, and planes which we represented on the diagrams which we used in our proofs. But the diagrams were certainly no part of the proof. They were suggestive and a useful aid to memory, but the validity of the proof was independent of the diagram. And the same is true of the commonsense notion of points and lines and planes which lay behind so many of our logical manipulations. They too can be eliminated. Instead of points, lines, and planes we might speak of ziggles, zaggles, and daggles, without any reference to the sorts of things which we denote by these words. And this lack of reference or meaning then applies not only to the entities of our mathematical systems—the points, lines, or planes—but also to the relationships between them, relationships like points lying on a line.

We might for example proceed to develop geometry as follows:[3] The entity which this subject considers shall consist of one set to be denoted by the lower case letters of the Roman alphabet, a, b, c . . . and a second set to be denoted by the capital letters of the alphabet, A, B, C . . . These two sorts of entities, we then say, may on occasion satisfy a relationship which we denote by the Greek letter ∈. Thus a ∈ A is a statement in our system. It will be a true statement for some *a* and some *A* and false for others. But we have no idea what it means. Yet in terms of these symbols we may set up postulates for our system.

One such postulate would run as follows. If two entities of the first sort are distinct, which we indicate by writing a ≠ b, and if in addition there are two entities of the second sort, A and B such that a ∈ A, b ∈ A, and also a ∈ B and b ∈ B, then it will always be true that A = B.

So far we have said absolutely nothing about what the little letters, the capital letters, and ∈ are. And we need not do so; we can continue to add other postulates without any such specification. We need only take care not to put in any postulates that contradict those we've already annunciated, and we must put in enough postulates so that the set of postulates as a whole will have consequences. Certainly systems of this sort can have consequences. For although you have no notion what these A's, B's, etc. are, you can readily see something which they might be. The lower case letters may denote the sort of thing we normally call a point; the capital letters may denote the sort of thing we call a straight line; and the Greek letter ∈ may be construed to denote the relation which subsists when a point lies on a line. Thus a ∈ A would read: the point *a* lies on the straight line *A*. With this interpretation the postulate which we have already annunciated becomes the one with which you are already familiar. If a and b are distinct points, and if both these points lie on the line A and also on the line B, then A and B must be the same line. Or, in the more usual phraseology, only one straight line can be drawn through two distinct points.

We see then that we may adopt this particular interpretation of the entirely abstract and undefined symbols which we have been using. If we do, we will get back to normal Euclidean plane geometry. But we do not have to adopt this interpretation or any other. All of the postulates can be written without any interpretation of the symbolic entities or of the relationship between them, and all of the theorems can be deduced from these postulates by logical operations like the substitution of equals for equals without any effort to clothe the symbols with meaning. Geometry thus becomes the study of the consequences under logical manipulation of those properties of the undefined entities and relations which are given in the postulate system. And all of mathematics can be developed in this manner.

We are thus introduced to the notion of mathematics as a purely formal study. Mathematics, we now say, has no content. It is not about anything. It is simply the study of the logical consequences of supposing that certain undefined entities satisfy certain relatively arbitrary restrictions when placed in certain undefined relationships. The necessity of mathematical truth is thus completely reduced to the necessity of logical truth, and the truths of mathematics like the truths of logic are devoid of consequence for our perceptions of the natural world.

But although mathematics need have no application to the world of our perceptions, it may have such application. We are at liberty to explore our perceptions in an effort to isolate entities and relations between these entities which will fit one of the mathematical postulational systems. We may, for example, specify a straight line in nature to be the path of a light ray, as we do when we test the straightness of a stick by sighting along the side of the stick. We may specify a point in nature as the intersection of two such lines, or by some other device. And we may determine, by experiment, whether the natural entities thus defined do in fact satisfy the postulates of our entirely abstract system. If they do, if these natural entities are capable of serving as an interpretation for the postulates, then the theorems apply to them as well. For the theorems are consequences of the postulates whatever the interpretation of the entities and relations of the system. And this holds not only for geometry, but for the numbers which we add and subtract and for our algebra. Our system of numbers and our normal algebra may be interpreted as applying to the distances which we measured with yardsticks or the weights we determine with scales, for the numbers shown on our yardsticks and scales, when these are applied to natural objects appear, as nearly as we can tell, to satisfy the postulates upon which our number system and our algebra are based. But our number system and our algebra need not be interpreted this way and their mathematical validity is independent of the interpretation.

Historically this newer understanding of mathematics has been of tremendous significance, for it has resulted in the realization that the particular postulational set upon which all mathematics before 1850 had been based are limited and arbitrary. There are many other sets of postulates whose consequences can be developed by logical manipulation, and the study of these other postulational sets has been extremely fruitful. We have developed a geometry in which the sum of the angles of a triangle is not equal to 180°, and we have developed algebras in which x times y is not equal to y times x. Many of the systems developed in this manner remain without physical interpretation, and their interest is in no way dependent upon the existence of such interpretations. But for other new postulational sets we have succeeded in finding physical interpretations. General relativity utilizes a geometry different from Euclid's; quantum mechanics utilizes an algebra in which a times b is not the same as b times a. And it is extremely doubtful whether these new physical discoveries about the world would have been possible without the prior understanding of the nature of mathematical abstraction. The physicist has profited greatly from the mathematician's discovery that since in mathematics one is not talking about anything one can provide the object discussed with almost any property one chooses. Recent history has made it abundantly clear that one of the fundamental roles in science of logical and mathematical formalism is the development of abstract postulational sets, without physical reference, but with formal consequence, which the physicist may attempt to apply to entities abstracted from the natural world. Theoretically the mathematician might wait and let the physicist provide him with a set of postulates abstracted from nature, and this has occasionally been done fruitfully. But more often it has turned out that the system required by the physicist had been developed earlier by the abstract mathematician who had no thought for its application. And the physicist's discovery of the possible abstraction from nature has normally been dependent upon at least a cursory acquaintance with the pre-existing branch of abstract mathematics.

But the application of logic in physical science is not restricted to its role in mathematics. The syllogism itself is not mathematical, yet much scientific reasoning may be cast in this form or better in the more generalized form provided by modern logic. So we must now investigate logic itself somewhat more closely.

Now it turns out that logic itself can also be formalized, that is, it can be reduced to a set of postulates completely without reference and to a set of statements about the manner in which the entities and relations appearing in these postulates are to be manipulated for the generation of logical theorems.

Perhaps an elementary illustration will make this clearer. In logic we may deal with a number of undefined entities denoted by symbols like p, q ... and we include as objects of our study a series of connectives by which the p's and q's can be compounded to yield new entities of the same sort.[4] We call the p's and q's propositions, and we gain new propositions from these with the aid of certain connectives. One such typical connective is the conjunction, which is denoted by a dot. If p is a proposition and q is a proposition, then p · q is another proposition.

These propositions can, we now say, have one of two values. They may be true, which we denote by T, or false, which we denote by F. By using this symbolism we can describe the properties of the dot connective completely with the aid of a table. In the first column we tabulate values of p, in the second values of q, and in the third values of p · q. The proposition p may be true or false, and for each of these values of p, q may be true or false, and the four possible combinations are indicated by the rows in the first two columns of the table. And the logical properties of the dot will now be completely exhausted if we specify in the third column the values which the proposition p · q will take for all the combinations of value of p and of q. This choice might be made completely arbitrarily but it is convenient and customary to place a T in the first row and an F in all the others. Thus if p is true and q is true, p · q is also true. But if either p or q or both is false, then p · q is false as well.[5]

p	q	p · q
T	T	T
T	F	F
F	T	F
F	F	F

With the aid of one more sign we can get a theorem. This sign is called the negation sign and is represented by the tilde. It does not connect propositions but changes one proposition to another, and its properties are completely represented by the very simple table which tabulates the values of p in one column and of ~p or not p in the other. We see that the effect of the tilde is simply to change the truth value of the proposition. If p is true, ~p is false, and vice versa. Now let us set up a similar table for the compound proposition p · ~p. Here we need only two rows. If p is true, ~p is false, if p is false ~p is true.

p	~p
T	F
F	T

Now we can look back to our old table for the dot connective and discover in both these cases p · ~p must be false. And this is a theorem: p · ~p is always false, or ~(p · ~p) is always true. It is valid by virtue of its form alone. It makes no difference what p is or what the value of p is, ~(p · ~p) always has the truth value denoted by T. And this is a consequence simply of the rules which we described in our table. Any theorem which can be developed in this manner is known as a tautology. There are many others of which this is simply the most trivial.

Not all logical truths are tautologies, nor can they all be derived with the aid of tables like the ones we have here considered. But all logical theorems share with our trivial example a number of important characteristics. None of them need be taken to be about anything. We have called the p's and q's propositions, but we might equally well have called them "snerlbarks." We have called their values true and false, they are also frequently denoted by 1 and 0, and any other signs might have been chosen. We have called the dot a conjunction and the tilde a negation, but we could as well have stuck to the names dot and tilde. We need not know what any of these things is, and we can still derive logical theorems. Or perhaps I should say we can still find combinations of these undefined symbols whose values are T for all values of the components. Thus we need not call this subject logic at all.

Viewed in this manner, logic, or whatever you wish to call it, is just a game, like chess. The game is played with certain undefined symbols or markers which may be interpreted in the case of chess to be kinds or queens, knights and pawns, but which need have no such interpretation, and these markers are manipulated according to certain set standard rules which we ourselves may select. We did not for example have to put one T and three F's in the column in which we interpreted the properties of the dot connective. We could have employed another set. And chess rules have the same characteristic. Chess does not have to be played according to the standard rules. Chess has been played in three dimensions, and it has been played in two dimensions on boards that are 8x9, or 8x10, or 10x10 rather than the normal 8x8. It has been played with extra pieces or with the normal pieces permitted to move in abnormal manner. And such variations can be multiplied almost without limit without destroying the basic pattern of the game.

But for all this freedom there are certain variations which cannot be permitted if the game is to be played at all. There must in the first place be some one set of rules fixed in advance of the game. You cannot change the rules as you go along. Nor, if the game is played on an 8x8 board can a piece be employed which is required to move through ten rows or ten columns at a time. Nor, to take an extreme example, may a chess piece be permitted to move from any square to any other square at will. With this modification white would always mate on the first move. And the same sorts of limitations apply to the rules of our formal game of logic. The rules themselves must be fixed in advance, and certain sorts of connectives must be barred. The dot, for example, would be useless if the table which described its properties contained a solid set of T's or a solid set of F's in the third column.

To this point I have suggested that the p's, q's, dots, and tildes, as well as the T and F of our logic need not be interpreted, that they are developed purely formally and without any reference. But of course we can find an interpretation for them. In fact, we have developed them in this particular manner in order that they might be capable of the interpretation which we are going to supply. For of course we may take these p's and q's, the propositions, to be the normal declarative sentences of our language. They may be everyday sentences like, "The apple is red," or they may be the symbolic sentences of mathematics like our older "a ∈ A," or a ≠ b, or A=B. Similarly, the dot conjunction may be interpreted as the "and" used to join simple declarative statements into compound declarative statements. From "the apple is red" and from "the sky is blue," this conjunction makes the compound declarative "The sky is blue and the apple is red." Again the tilde or negation sign supplies the function of the *not* in everyday language. From "The apple is red" the tilde makes "The apple is not red" and so on for the other undefined symbols which a fuller treatment of logic would develop.[6]

So far the identification is arbitrary, but it turns out that it can be usefully extended. For if we now identify the T and the F of our table with the true and false of normal discourse, then the statements that our table shows to be tautologously true are among those general forms or compound statements of ordinary discourse whose validity or falsity is uniformly granted. Thus in p · ~p always being false, we see a symbolic summary of the universally admitted truth that compound statements like "the apple is red and the apple is not red" or "the point lies on the line and the point does not lie on the line" are always false. And again this procedure can be extended to include those other linguistic compounds whose validity or falsity seem everywhere to be admitted.[7]

It was certainly not clear in advance that these universally acknowledged truths, like the syllogism, could be reduced to a pure formalism, to a game played according to relatively arbitrary but completely binding rules. And the possibility of such a reduction provides an important clue to the apparent necessity of all such logical truths. For the rules of our formalism are now seen to correspond to rules implicit in our language. They represent relatively closely certain of our linguistic conventions, and if complete they should include all those conventions which are independent of the meanings of the words employed in language and which derive from the form in which these words are arranged for communication. And the impossibility of doubting the validity of the syllogism or of some other logical derivation presumably derives not from any particular characteristic of the external world or of our minds, but rather from a set of conventions which we have adopted in order to be able to talk about the world, in order to be able to communicate at all.

The rules of formal logic or of formal language are conventions. They are the rules of the game which we play with other human beings when we communicate. In themselves, they are no more necessary than any other adequate set of rules, but without some such set no communication or very little communication would be possible. We are not then obliged to admit the necessity of the syllogism, but an announcement that we will not do so is an announcement that we will not bide by the rules of the game, that we will not play. It is therefore in the most literal sense antisocial, it carries the penalty of other antisocial acts, and it deprives us of the privilege of learning certain things from the experience of others.

I confess that this view that the truths of logic are products of linguistic convention is not by any means free of difficulty. It is probably true that it raises as many questions as it answers, and the problems raised have led a number of philosophers and logicians to reject it. I adhere to it at this time in spite of my inability to resolve all the difficulties simply because the difficulties seem not more severe than those which arise in the attempt to root logical necessity in the external world or in the innate categories of the mind. Yet the difficulties cannot be simply dismissed, for they indicate directions in which further clarification must be obtained.

In the first place, the methods of formalizing logic which we have discussed are circular. In setting up our truth tables and describing their utilization we required language. Yet we suggested that the regularities embodied in the tables underlie all our language. And we should therefore have liked to suppose that the formalism was prior to the language for which it legislates, that it could be constructed independently. But although it is reasonably clear that this circularity cannot be eliminated, it is by no means so clear that it constitutes a real difficulty; or just what the difficulty is if it exists. We have already noted that the procedure of cutting up the world in science or in everyday life is a circular process through which increased precision of the cut is obtained, and there is no apparent reason why the same should not be true of linguistic convention.

A further difficulty may reside in our use of the word linguistic. It may not be either necessary or desirable to associate these conventions with purely verbal communications. There are other more primitive forms of communication, like the gesture, and these too may have their own conventions and therefore their own logic. It may further be suggested that our association of logical convention with communication is itself an error since the conventions seem requisite to thought even though the thought is not communicated. But this consideration involves us with the extremely controversial question as to whether thought itself is possible in the absence of a socially forged symbol-system. And today it appears impossible to say more than that the two are surely not independent.

Far more important than either of these difficulties is our present lack of understanding of the extent of our freedom in the choice of the conventions which underlie logic and language. Are our conventions the only ones? Is our logic the only logic? Or might we abstract other conventions if we considered languages radically different from our own? To what extent is the structure of Indo-European language responsible for the highly developed state of our mathematics and our science? Such questions suggest the further possibility that our logical conventions themselves as well as the entities to which we apply them may be related to the manner in which we form the behavioral worlds discussed during the last lecture. And this, if true, would necessitate a qualification of our initial remark of the independence of logic and experience. Logic would remain independent of particular experience. Logical truths cannot be invalidated by experience. But the logical conventions might again be involved in the categories to which we subject our experience so that if the conventions are not unique, changes in them might be associated with changes in the behavioral world and therefore with changes in experience.

These speculations need answers, but our uncertainty about the answers does not bar the possibility—or to my mind the probability—that logical and mathematical truths are grounded in conventions. And these conventions, though more or less arbitrary in their particular form, are in some form necessary preconditions of communication, of learning, and perhaps of thought itself. The recognition of the existence of such conventions has led to a persistent proposal that we reduce all language, or at least all scientific language, which some philosophers take to be the same thing, to a purely formal system supplemented by a set of rules determining the meanings in the perceptual world of the undefined elements of the formalism.

The nature of this proposal for the formalization of scientific language will be clearer if we revert to our earlier example of geometry. There you will remember we dealt with a group of entities with no reference, our a's and our A's, and with certain relations between them among which was the one which we denoted by $\in$. These entities and relations had no referents, but they did have certain properties which were completely described by a set of formal postulates, and there were a number of different pure postulational systems which could be prescribed for them. Further, these postulational sets had consequences derived from them by considering $a \in A$ and similar compounds to be propositions, to be the p's and q's of formal logic. And applying to these compounds the conventional logical manipulations, the formal consequences of the postulational sets were discovered.

It then turned out that the a and the A could be interpreted as the point and the line of a variety of different abstract geometries. One of these geometries was Euclidean, others were non-Euclidean. But among the many geometries generated by different postulational sets was one which we can call the geometry of actual space. And we discover which of the geometries this is by equating straight lines with the path of a light ray, a thing we determine by experimental manipulation and by equating distance in geometry with the result of a series of operations performed with a yardstick. That is to say, from among all the possible abstract geometries constructed by applying the conventions of logic to postulate systems involving the a, the ∈, and the A and other referenceless symbols, we select one which can be applied to the real world. And the criterion determining this applicability is that for this geometry we can describe a set of manual operations which determine entities and relations like straight lines and distance in such a manner that these entities and relations will satisfy the postulate set or at least will satisfy it arbitrarily closely.

This geometry is then the geometry of real space. The theorems in it apply to real space, and we have found physical referents for the previously unattached symbols of the abstract mathematical system. In this way, our knowledge of the space of the world, of the space provided by our perceptions, is reduced to an understanding of a purely conventional formalism and of the consequences of this formalism, together with an understanding of a set of operations by which the abstract entities of the formal system are given reference to entities abstracted from the perceptual world.

The program for the formalization of scientific language calls for the application of this technique to other portions of science and to the language of scientific communication. And scientific language is for this purpose taken to be the entire language in which we discuss our perceptions. In everyday terms, it is the language in which we discuss facts or events or the relations between events. It is a language used not only by the scientist but, more loosely, by the layman. And it is this language which we are now asked to formalize.

The general direction in which such a program should proceed ought now be reasonably clear. We must introduce, in addition to the a's and the A's which we may interpret as points and lines, additional abstract symbols and arrangements of these symbols which we intend later to be able to interpret as temperatures, forces, masses, and so on. Just how many such symbols we require is not clear, but we must have enough for the complexity of our perceptions. We next set up pure formal relations between these abstract entities, and we investigate by logical convention the consequences of our postulates.[8] This is the purely formal component of our scientific language.

But we require in addition a set of rules enabling us through our perceptions and our manipulations to correlate these abstract entities of the formal system with entities abstracted from the world of our perceptions. These rules determine the meanings of the formal entities of the abstract language. The application of the rules determines in the perceptual world a group of entities and relations satisfying an appropriately chosen postulate set; and these entities of the perceptual world are the referents of the abstract entities of the formal system. So all scientific language, all science is reduced to convention and its consequences plus meaning-relations. And the ultimate outcome of this program would be to separate entirely the empirical portion of our knowledge from the portion due to the conventions we employ in gaining and formalizing it.

It is by no means clear how far this project can be carried. The efforts in this direction are still in their preliminary stages, and it is far too early to reach any final judgment upon them or upon the projects from which they derive. But certain preliminary remarks are surely called for. I am myself considerably less optimistic about the possibility of completing this research effort than the majority of contemporary students. And I shall dwell at some length, both this evening and in the early portions of our next lecture, on what I consider to be the limitations of a completely formalized language. But first we should clearly recognize that whatever the ultimate limitations of the effort, the attempt to formalize limited areas of existing scientific knowledge has been and will continue to be a useful tool of research.

We have already noted the manner in which physics profited by the formalization of geometry. Even while incomplete, this formalization resulted in the recognition of the existence of many geometries different from Euclid's, and this recognition increased the freedom of scientific thought about the characteristics of physical space. Here clearly is an example of a scientific situation in which the separation of empirical and formal elements of knowledge provided the scientist with increased freedom of conceptualization. For the Euclidean formalism which had previously been thought necessary was shown to be only one of a number of alternate formalisms, and the scientist was directed to new aspects of experience in an effort to choose among possible alternates.

The history of science is filled with examples of the fruitful application of logical formalism to mental concepts like those of matter, of motion, and of space. Logical necessity and empirical necessity are all too frequently confused. Their separation normally, as in this this case, results in the recognition of new possible formal categories and directs experiments to new aspects of the perceptual world. We have already noted the important effect of successively more adequate logical solutions of the Parmenidean problem upon the science of antiquity and of the seventeenth century. If we had the time we might equally well have examined the effect in the nineteenth century of the formal analyses of the Newtonian laws of motion, for these together with the analysis of space conducted during the same period led once again to the recognition that the empirical basis of Newtonian mechanics could be compatible with other formalisms, and this recognition is an important portion of the intellectual background of our radically differing modern physics.

In these examples we see the application of formalism to a limited area of knowledge as a tool which by creating greater conceptual freedom aids in the resolution of crisis states in individual sciences. But the application of logical formalism can also create such crises. We discussed an example of this sort in considering the scholastic contributions to Galileo's reformulation of the problem of motion. One part of this work, you remember, was completely independent of any observational material, old or new. It was a study of the logical categories adequate to the consideration of any change, and we might paraphrase the outcome of this study by saying that it led to the recognition of the logical or formal incompatibility of the existing concept of motion with the formal categories applied to the analysis of motion. Conceptually the motion was grasped as a whole; analytically it was equated with its endpoints. And, as we noted again in studying the child's view of the problem of motion, these two approaches lead to conflicting formalisms.

So formalization can be a source of scientific crisis or it can be an aid in the resolution of such crisis. And it has one even more important function. It conditions the structure of our texts, the manner in which we organize scientific knowledge, and the manner in which we transmit it. The structure of a text, as we discussed it in our first lecture, was a formal structure. The laws were stated in abstract, possibly mathematical terms; formal deductive techniques were then described; and finally a set of operations was proposed by which the formal consequences of the law could be compared with experience. These operations constituted the meaning system of the formalism.

Of course this description is an idealization. No existing body of scientific knowledge has as yet been reduced to a formalism paralleling in its completeness or abstractness the formalism to which we have reduced most of mathematics. Nevertheless it appears likely in principle that such formalization can be achieved. But whether the meaning system for such a formalism could be adequately described is less clear. The modern philosophy of science known as operationalism bases much of its program upon the supposition that it can be. But no adequate illustrations of such a scientific meaning system have yet been provided. To say as we have up to this point that distance is simply the result of an operation performed with a yardstick is in itself inadequate in precision. Certainly one must also describe the material of the yardstick, the temperature at which it is to be maintained, and the manner in which it is to be moved if the length to be measured is greater than the length of the stick itself.[9] The failure to prescribe some such unique operation for the determination of distance makes synonyms of various different operations used to define the single term *distance*. And this is equivalent to making certain sorts of synonymy depend upon previous scientific experience which is surely not a satisfactory characteristic of a meaning system. But the alternative provided by the unique definition is too narrow for practical application. Certainly our texts do not even attempt to approximate such an ideal.

So it is possible that our texts, including meaning systems, cannot be fully formalized, but they can and do approximate however inadequately this ideal. And even so limited an approximation is useful. It permits a tremendous condensation of the content of scientific knowledge and, by separating even partially the formal and empirical elements of knowledge, it facilitates verification of the end-products of scientific conceptualization. The condensation of the empirical content of science is, if you will, simply a mnemonic device,[10] but from it proceeds much of the cumulative character of scientific knowledge. The separation of the logical from the empirical elements in the text increases the freedom of scientific conceptualization by partially eliminating the effect of the orientation from which the theory was derived and thus makes easier further advances in scientific theory. But, as we noted, there is real question how far even this separation can be carried.

All of these examples of the fruitful application of formalism have one important characteristic in common. In all cases formalism was applied to an existing organized body of scientific information and scientific theory. But the proposal for the construction of a scientific language greatly transcends these limited applications. Basically it is a proposal that not only the products of research but the research itself be designed and expressed in terms of a formal language plus a meaning system. Fundamentally, it calls for the construction of a language adequate to deal unambiguously with infinite variety presented by the totality of our perceptual experience. And this I think is clearly impossible.

At the start of the next lecture I shall discuss the effects of the completion of such a program upon the language of everyday discourse. For we can learn a great deal about the limitations of formalism from such an elementary discussion. At that time I will suggest that such a language would either be infinitely complex and therefore totally unmanageable, or alternatively that it would be incapable of dealing with *new* experience. But in the few minutes remaining this evening let me simply suggest the difficulty of reconciling such a program with what we have already learned about the nature of the scientific process.

If one were to commence such a project of complete formalization, one would undoubtedly start with the language employed for existing scientific knowledge. Perhaps the program could be completed; certainly existing textbooks provide a preliminary step in this direction. But supposing the project completed, its results would be to freeze scientific attention upon just those aspects of nature which are embraced by contemporary science. It would provide a place in its meaning system for aspects of nature now considered technically relevant and no place for others. As a result it would not be a language adequate to embrace new conceptual developments in science. If the new language were developed to fit present-day science we should have to change it with every change in scientific theory. For we cannot tell in advance what aspects of nature will prove important to our forthcoming scientific theories. So we require a broader language than the language of contemporary science. We need a language broad enough to embrace aspects of experience not now considered scientifically relevant. Certainly we can make the language somewhat broader, but our knowledge of behavioral worlds, discussed in the last lecture, may make us wonder whether we can broaden it enough. For we have reason to suppose that our perceptions themselves simplify the total perceptual flux, and that this simplification can only be altered when and if it proves behaviorally inadequate. And surely unless our ability to discriminate among our perceptions is fixed once and for all, we cannot design a stable language to deal with them all. A formal language adequate to deal with changes in the behavioral world would demand an advance knowledge of the totality of all our future perceptions, and we have every reason to suppose that there is no basis, theoretical or empirical, upon which such knowledge could be grounded.[11] Yet the scientist himself lives in a behavioral world. Changes in scientific theory are associated with changes in the behavioral world of the scientific profession. And they are therefore presumably associated with changes in perception with which no fixed language can deal.

To this point we shall return in the final lecture.

Notes

1. Kuhn noted to himself here, and for the example in the next paragraph, "Put on Board." "Barbara" is one of several mnemonic names for different forms of deductive syllogism. Here the three a's in "Barbara" represent the three quantifiers "all," as in: "All Greeks are human," "All humans are mortal," therefore "All Greeks are mortal," of which Kuhn's version about the singular Socrates is an adaptation.

2. Kuhn instructed himself to put these formulas on the blackboard.

3. Kuhn instructed himself to put these formulas on the blackboard.

4. Here again Kuhn directed himself to the blackboard.

5. Kuhn's scripts do not include the following two truth tables, which I have inserted for illustration.

6. Here Kuhn inserted by hand, "set up if time" the following formula:

$$\text{If } (a \neq b \cdot a \in A \cdot b \in B \cdot b \in A \cdot b \in B) \text{ then } (A=B)$$

He labeled the arguments and conclusion, respectively, (p, q, r, s, t, and u). Kuhn had discussed this argument previously in the lecture.

7. Here Kuhn's text reads, "Among these would be the syllogism Barbara, which in its full linguistic form would be exemplified by 'If Socrates is a man and all men are mortal, then Socrates is mortal'." The first six words, "Among these would be the syllogism" are typed and the rest in handwritten. Kuhn may have forgotten that he introduced the syllogism form Barbara at the beginning of this lecture. Accordingly, I have removed this second introduction from the main text.

8. Kuhn parenthetically typed, but then crossed out, "Insert remark about introducing sufficient symbols to give us all words of our language."

9. Next to this discussion of operationalism, Kuhn noted in the margin, "Expand if possible."

10. Here Kuhn inserted by hand "we need state only postulates."

11. Kuhn bracketed this sentence, possibly for omission.

Lecture VIII

Canons of Constructive Research

I should like to continue this evening the discussion begun in our last lecture of the difficulties inherent in the formulation of a completely formalized language for all of scientific research. Some of these difficulties are displayed particularly strikingly in a discussion of the effect of formalism upon natural language, the language which we use every day in describing what goes on about us. I do not of course mean to suggest that scientific language and natural language are synonymous. Scientists alter and refine the common concepts and words they use, and they introduce new words and concepts peculiar to their own disciplines. But as we have noted again and again, natural language and everyday concepts provide the starting point for all scientific investigation. If scientific research can profit by the complete formalization of language, then this formalization would have to affect natural language as well. And it will, I believe, be clear in what follows that the particular problem of linguistic formalism which we are now to consider must necessarily carry over from natural language to any more scientific which shares with it the task of mediating past and future perceptions.

One *sine qua non* of any formalized language is a precise and unambiguous meaning system. For every symbol of the formalized language there must be given a set of rules implicit or explicit which uniquely determine the perceptual complexes which this symbol correctly denotes. This set of rules prescribes, or perhaps itself constitutes, the meaning in the perceptual world of the symbol or word in question.

Now there are in theory two distinct manners of setting up such a meaning system. The first of these might well be called definition by enumeration or by pointing. We point to one or more objects in the external world and recite simultaneously the phrase, "These objects and only these objects shall be denoted by such and such a symbol or word." The best example of this procedure in practice is the allocation of proper names. "You Tarzan, me Jane," said a famous heroine to her new-found jungle friend. And the name or label stuck.

But there is another and more usual manner of providing a meaning system. The meaning of a term may be prescribed by the description of a set of criteria which uniquely determine the applicability of a given symbol or group of symbols to a given perceptual entity. Thus, a bachelor is a man who has not been through any one of a set of procedures which result in matrimony, or, who having gone through such a set, has followed it with another set which culminates in divorce. With a fuller description of these procedures, we are in a position to determine whether or not a given individual is a bachelor. We may interrogate him, or if this is unsatisfactory, we may examine legal records.

It is quite clear that at least in theory these two methods are entirely distinct. For words which are, by enumeration or pointing, applied to the same perceptual complex may be so applied by virtue of quite different criteria. Thus descriptive phrases like "the author of WAVERLY" and "the author of IVANHOE" refer to the same entity, but they are applied to it by virtue of different criteria. It is usually said that these two phrases refer to the same individual or that they have the same "referent," but that they have different meanings. The term "meaning" is thus reserved for the criteria which determine the applicability of the term in question, and it is observed that synonymy of reference and synonymy of meaning are two quite different relationships.

So two different methods of setting up meaning or reference systems are distinct. Either of them, taken alone, is capable of providing a systematic relationship between a formalized language and the situations to which it applies, but their application results in two rather different ways of cutting up the world. In practice we appear to use a combination of both.

These two methods are, so far as I know, the only two capable of providing precise meaning or reference for those symbols of daily discourse which we normally call nouns. Yet I do not believe that either singly or in combination they could be employed quite as described in the construction of a language capable of adequately mediating the entire world of our perceptions. This is particularly obvious as regards the first, the one which we called definition by enumeration. For there is not normally sufficient time to permit enumeration. To discover how to use the word "elephant" we cannot go about and point to all the elephants in the world. By the time we had finished our first trip there would be more elephants. We should never be able to use the word. Apparently for words like this we must rely on some description of the criteria according to which the word is to be applied. We may learn of these criteria by experience with a few objects to which the word applies. But we must learn such criteria for the word may have to be applied to a particular individual which we have not already seen.

Definitions by enumeration or pointing are possible only in the case of proper names, and even here a real difficulty is present. The infant to whom we give the name John Smith changes as he grows older, so we would seem to require some criterion beyond the initial pointing if we are to continue to call him by the same name. So finally definition by enumeration seems useful only in a world, or in a part of a world, which contains a finite number of entities and in which we can be perfectly certain of the perceptual stability of these entities. Such a stable universe, occupied by a finite number of distinct and perduring entities, is surely not the sort of universe mediated by our natural language or by the language of science.

The limitation of the alternate means of prescribing a meaning system are less apparent, and we will more profitably approach them by discussing the manner in which we actually arrive at meanings and then considering the manner in which the nature and function of language would be changed by an adherence to a more precise meaning system. We learn words in a number of different ways. We learn them from dictionaries and from other people. We learn them by observing the situations in which other people use particular words and as children we learn at least some of our words by random experimentation with sounds accompanied by the observation of the effect on other people of particular sounds. And of these procedures the last two are more basic. We learn our basic vocabulary through use, our own use or other people's use. So it is this acquisition of meaning through use which I should particularly like to consider.

How, for example, do we acquire our notion of the meaning of the word "dog"? We see pictures of a number of dogs, we watch as our parents or our friends point to objects to which they apply the word, and we gradually learn how to correlate the word with our perceptions. Our first notion may be very crude—we may apply it indiscriminately to cats, dogs, and horses. But gradually we learn to discriminate linguistically among these elements of our behavioral world. Certainly our linguistic discriminations become more precise; probably there is an accompanying increase in the precision of our perceptions. Gradually we have categorized the world; we have cut it up into boxes containing dogs, horses, cats, and so on.

You will recognize that this is an extremely crude heuristic procedure. But for most and perhaps for all purposes of everyday life it proves entirely adequate. We do learn in this manner to use a large number of words in the same way that other people use them under the same circumstances, and this is all that's required for purposes of social communication.

But obviously this manner of learning words does not lead to anything that could properly be called a precise meaning system in either of the senses of meaning that we have already discussed. We may repeatedly use the word dog correctly with respect to particular animals which we have never seen before. But this does not imply that we can produce or demand a set of criteria which govern the applicability of the word to particular perceptual complexes. If you examine your own concept of dog you will, I think, find that you apply it and use it without any corresponding knowledge of the criteria which make it applicable. So that if meaning is to be equated with an understanding of a meaning system, then many of the words of daily discourse can be and are applied without our knowing what they mean.

Our lack of a clear understanding of the precise meanings of the words we use has many disadvantages which we will want to discuss. But first it would be well to be entirely clear as to the difficulty which will confront us in any effort to remedy the situation. For although relatively clear criteria of applicability can be provided for certain of the words of natural language, there are others like "dog" with which it is peculiarly difficult to deal.

It is quite easy to provide a criterion which will distinguish horses and dogs. Horses have hooves, dogs have padded feet and claws. But the distinction between dogs and cats is not so easy to make. Some lap dogs have external appearances very much like those of certain breeds of cats. Some wild cats are very near in their appearance to some of the larger breeds of dogs. But again a criterion can be found. Cats have retractile claws, dogs cannot retract their claws. This is inconvenient as a criterion. No one wants to examine the claws of strange animals. But it will serve.

However, when we come to the attempt to distinguish dogs from animals like wolves, foxes, and jackals we have an extremely difficult time. Here we appear to be dealing with members of the same biological genus. Neat distinguishing features like the one based upon types of claws are not available to us. Instead we are likely to resort to a discussion of the normal habitat of these various breeds, of the possibility of domesticating them, or their normal food, and so on.

But these criteria are different from the ones that we've imposed so far, for these, though quite good enough for the purpose of normal discrimination, might very well be violated. Wolves have occasionally been domesticated. They then act very much like certain sorts of dogs. Besides wolves and dogs can crossbreed, and we can by judicious interbreeding produce a virtually continuous spectrum of animals from the wild wolf to the domestic dog. And then where will we draw the line between them? And to this question the answer is obvious. We may draw it anywhere we please. It could not make less difference. It is not one of the jobs of everyday language to distinguish between these cases that might occur, but are supremely unlikely to occur. So after all this effort we may well conclude that the job was not worthwhile. The vague definitions implicit in our usage of words are normally just about as precise as they need be for our practice.

So it is difficult to describe any precise meaning for the word dog. I have used the word frequently. So far as I know I have always used it correctly, but I had to go to the *Encyclopedia Britannica* in order to learn even the little bit about dogs which was requisite to the above attempt at definition. And even with that knowledge we fell short of complete precision. Nevertheless at the expense of being occasionally arbitrary, we could achieve such precision.[1] We may illustrate our precise definition with the aid of a diagram. Here is a rough circle. At its interior we say lie all animals which are four-legged and have non-retractile claws. Thus the area included within the circle includes all dogs, but it also includes some other animals—wolves, jackals, etc., and we want to get rid of these. To do this we provide the most precise description possible of the animal and we draw little circles representing these descriptions. These animals are to be removed from the class of four-footed animals with nonretractable claws. The remaining area provides our precise definition of dogs.

Of course in providing the descriptions by which we drew little circles we may have made some mistakes. We may, for example, have drawn these little circles in such a way that the offspring of two animals which clearly fell in the class of dogs falls in the class of wolves and is not therefore a dog. To this extent we have been arbitrary. These are the cases that are not going to make any practical difference. We can choose to apply the word as we please, and we have made one particular choice.

In this way we've gotten a precise definition. But we now have to recognize that the precise definition is relatively useless to us. In itself it will not supply the function of what we normally take to be the meaning of the word dog, for by dog we mean a great deal more than this, and we do not mean quite this. For we do not look at claws. We apply the word by other criteria. Dogs are things which bark and frequently bite. They are good household pets, man's best friend next to his mother, and they are clever—they can be trained to do tricks and perform useful functions. All of these things are involved in what we ordinarily mean by dog. If enough of the expected characteristics apply then we use the word. So, to get back from our precise definition to our normal concept we will have to add to the definition a group of laws about dogs, and we can list some of these. "Dogs bark"; "dogs bite"; "dogs can be domesticated." But this one gives us some difficulty: perhaps some dogs can't be domesticated. We'd better substitute, "most dogs can be domesticated." As another law, we may write "dogs are fur-bearing animals." Ah, but this is surely wrong. There is at least one breed known as the Mexican hairless which is not fur-bearing. So we must write this, "All dogs except . . . are fur-bearing" and for the ellipses we substitute a description of the Mexican hairless. And we can extend this list at some length.

But already you see what is happening. Normally our concept of dogs contains all these elements together though not necessarily explicitly or precisely. Our concept is thus vague. We can make our notion of dog precise, but in so doing we rob it of most of its pragmatic value. The application of the precise word "dog" tells us very little about the thing to which it is applied. In order to substitute for the knowledge which we have sacrificed in making the word precise, we must annunciate empirical generalizations about the class of entities embraced by our now precise definition. But as soon as we start such a precise annunciation, the vagueness of the old concept reappears as vagueness and lack of precision in the generalization. Our knowledge, our experience is simply not adequate to permit us to be precise about the two simultaneously. The vagueness which first appeared in our description was an index of ignorance, not of sloppiness.

Actually this distinction between the definition and the empirical generalization about the defined entities is highly artificial. We do not even approximate such a situation in our use of natural language. There is no one characteristic or set of characteristics by virtue of which we apply a name. We do not say "individual X is defined as the author of WAVERLY, and empirically we know that he is also the author of IVANHOE," any more than we say "individual X is by definition the author of IVANHOE and we know by experience that he also wrote WAVERLY." Theoretically we could define the individual in either way. Actually we defined him neither way or both ways. In our definition the arbitrary and the experiential are inextricably intermingled.

This is seen even more clearly as regards our example "dog." Certainly if we want to know whether a given entity should be called a dog, we do not look first to see whether the claws are retractile. We are far more likely to look first at certain of the characteristics predicated of dogs by our general laws. Roughly we may describe our concept according to the following diagram. At its center is what we may call our hard core of meaning, the attributes of which we are relatively certain. Dogs bark, they bite, they have four legs and nonretractile claws. In a ring outside of this center come the attributes of which we are relatively certain: dogs are fur-bearing for the most part, dogs are normally tameable. You notice that I include here as part of the meaning of dog "fur-bearing," which I know is occasionally violated. But it is still usually a useful way of judging whether what I see is a dog. I simply must be prepared to be mistaken in using this criterion alone, and I won't use it alone. In still a third and larger circle we include a number of attributes about which we are even less certain. For example, our expectations that dogs are actually tame, that they can be trained, that they have a name to which they will respond.

That we actually do use words in some such way contributes a number of important characteristics to language. In the first place, although by social usage and the necessities of behavior there will normally be a very considerable measure of agreement about what I've called the hard central core of meaning of a particular word, there may be a very large measure of disagreement about the vaguer fringe of meaning. This disagreement may arise from divergent previous experience, differing professional needs, or from many other factors. Its result is that many people who have always used a particular word in the same way may have considerable disagreement as to what the word really means. They may agree uniformly about its application to objects that they've experienced up to the present time, they may disagree about its meaning, and they will therefore disagree about its application to a new object of experience which they may not yet have met.

Thus we can have violent arguments about matters of definition, even though definitions understood in another sense are entirely arbitrary. And these arguments about definitions, which are quite independent of the fact that we apply the given word in the same way, are important because the particular definition to which we adhere will determine in some future case how we act in a new situation. We may argue violently as to whether a hypothetical animal which looks exactly like a wire-haired terrier but which has retractable claws ought to be called a dog or a cat. But if we meet such an animal we will behave somewhat differently depending upon which side of this question we are committed to. This example is of course trivial for we are virtually certain that we will never experience such an animal. But the history of biological classification demonstrates that not all such cases are trivial. New experiences do force us to change the meaning of older words, and the debates which arise at times of shifts of meaning are in large measure products of the differences in the criteria which we have previously considered most central to the applicability of the term. So arguments about definition are possible and they are important, even for words which are applied in the same way by everyone. For it is only through such arguments that we can make our definitions uniform. And differences in definition will affect future behavior.

It must be obvious by now that these vague meanings, in which are embodied much of our experience and many of our expectations as to future experience, would be of no use at all in a world other than the one we live in. We can employ our vague notions of dog only because in practice the world of our perception does not contain a complete spectrum of animals. There are no intermediates between the dog and the cat as regard certain distinguishable characteristics or between the dog and the horse as regards others. The concept of the dog represents a cut or a box in the perceptual world and its boundaries may remain relatively vague because the perceptual world presents us with few or no entities lying near the boundary. If the world were continuously populated, if all conceivable entities existed, our language system would be totally useless.

But although the world is such to permit our using language in this way, and although this is the way we actually use it, it would be foolish to pretend that there are not grave dangers involved in the use of the vague sort of meaning system we have here described. Such vagueness of meaning makes us extremely liable to errors of prediction. We see something we take to be a dog. With dog we associate a relative tameness, so we venture closer. But this turns out to be an undersized grizzly bear, and we lose an arm. Not infrequently we read of children who have made this mistake and lost their lives. It is possible to say that our mistake in this case is not a mistake in the use of words but is a mistake in the interpretation of what we have seen. But this was the terminology we examined and rejected in our discussion of behavioral worlds. Our words and our meaning system, like our senses, discriminate perceptual entities primarily with respect to the differences in behavior called for by the various entities. Dogs call forth one sort of behavior, bears another, and these behaviors are implicit in the meanings which in practice we give to these words. If we lose an arm to a bear thinking it a dog we change the meaning of "dog" to avoid the repetition of this mistake.

The dangers inherent in the loose use of language are of course even more severe in our political and social life. It is against these dangers that we have been so frequently warned recently in writings on the tyranny of words.[2] Words like radical, reactionary, or communist are encumbered with all sorts of associations whose conjunction has little reference to our experience. We label a man communist because of one relevant or irrelevant aspect of his behavior, and we then assume that he possesses all the other attributes which we associate with the name. And we do ourselves grave injury in this manner.[3]

The presence of such dangers in our use of language has led to an increased recognition of the necessity for great care and responsibility in our use of words. Words, as recent history has shown, are weapons. We must exercise extreme caution in manipulating them. But there is also a large body of opinion that insists that responsibility alone is not enough, that the only way we can really get away from these dangers is to demand absolute precision in our use of words or to employ only formalized languages with rigid and unambiguous meaning systems. It was of course languages of this sort that we discussed during our last lecture.

Now this I think is an impossible demand to impose upon any language which is to serve us in everyday life, or for that matter in scientific research. We can of course design such a language, but in the course of doing so we necessarily deprive them of their utility. And there are a number of reasons for this.

The first of these we have already hinted at in our discussion of the precise definition of a word like "dog." In making it precise we emasculated it. We deprived it of those components of its meaning which determined our reactions toward the entities with respect to which we employ it. These former components of meaning then had to reappear as generalizations about the entity dog now precisely defined, and as soon as we tried to state these generalizations, the vagueness which had formerly been inherent in the meaning of the word appeared in our generalization. We had therefore gained nothing by making our definition precise. To get rid of this vagueness, we should have had to reject all the generalizations except those about which we were absolutely certain. But to reject these somewhat uncertain generalizations would have been to deprive ourselves of a set of heuristic rules which, however inadequate they may be, do determine our activities. We cannot wait for the final experimental test in order to act. If you see a bear, you'd better run away. You can't afford to wait to see if by some chance it may turn out to be a dog.

This suggests a sense in which we can gain nothing by increasing the precision of our definitions, assuming of course that we have been responsible in applying experience to the determination of our original vague definitions. But I would suggest that the situation is even worse than this. Beyond a certain point an increase in precision is actually harmful. And this can be seen in the following fashion.

We have already seen that our language represents a manner of cutting up the world. It is closely associated with our cutting up of the behavioral world which we discussed in the sixth lecture. And among the boxes at which we arrived in cutting up the behavioral world is the one which contains dogs.[4] Now let us once again employ a diagram. Suppose that the entire area of this blackboard represents that portion of our perceptions which we would describe as perceptions of four-legged animals. It is thus part of the total universe composed of all things we might perceive. To represent the balance of this universe we should have to add a number of other boards, one for two-legged animals, one for plants, another for artifacts, and so on almost indefinitely.

Now we may proceed to divide further that portion of our perceptual world containing four-legged animals. At one point we draw a circle containing the particular combination of qualities and behaviors to which we expect to apply the word "dog." This circle is of course not precisely defined. Near its center is the hard core of meaning representing the attributes like nonretractile claws, meat-eating, and barking which we feel perfectly certain are associated with dogs. Nearer the periphery are attributes like tamed or tameable about which we are somewhat less certain, and of course the attributes marked in this peripheral area will display variations from one individual to another.

Now, our perceptions of four-legged animals contain more than just this one box. We'll have a similar somewhat imprecisely defined box for our perceptions of cats, another for cows, another for horses, still another for elephants. We will also have a box for wolves which in practice displays a slight overlap with the box for dogs. This indicates that under certain conditions we do not know how to tell a dog from a wolf if our particular perception is of an object lying in the fringe area of vague meaning. But generally there will be few such overlaps. In fact the totality of boxes which we erect in the area of our perceptions corresponding to four-legged animals by no means fills up the area of possible perceptions of four-legged animals. We have no name for an animal that is half-dog and half-horse, for we don't expect to see any such thing and do not take it into behavioral account. So we may say that the world of our possible perceptions contains some entities which we name with certainty, some for which we have no name and no corresponding behavior, and some intermediate perceptions about whose names there is disagreement and to which we should apply names only tentatively.

There are an infinite number of precise cuts with which we may replace this vague meaning system implicit in our natural language. Let me start by suggesting two extreme cases. We might begin by restricting meaning to those aspects of perception with respect to which we had absolute certainty. This would be equivalent to stating that the meaning of the word "dog" is to include just those perceptions of dogs which we have already had, and so forth for cats and other animals. And this means in effect that we narrow each of these circles representing meaning to some small portion of the hard core of meaning located near their center. This supplies us with complete precision in our use of language but it gives us a language applicable only to the perceptions which we have already had or to those future perceptions identical with these. It leaves the bulk of the world of possible perceptions without names, and we are bound to encounter such new perceptions which our narrowly precise language system will be unable to mediate. In gaining precision by removing the vague areas of meaning from our concept this language has lost much of its utility for us. It can be applied only to what we have already experienced.

But we need not attain precision by narrowing the meaning; we are equally at liberty to broaden it. In this case we arbitrarily eliminate the overlap in our diagram, we extend the outer rings on our circles so that they fill up the entire board without overlap, and we insist that any complex of perceptions shall be given the name corresponding to the box into which it falls. And there will always be such a box for it.

As compared with our previous cut, this alternative gives us the great advantage that it supplies the name for any perception, but it has an equally disastrous consequence. The meanings associated with these new precise cuts are so broad as to be behaviorally almost without significance. For in broadening the meaning of dog so that it includes not only the particular perceptual complexes which we have already experienced but also a large number of others which we have not experienced and probably never will experience, we have deprived the label dog of just that characteristic which made it a useful component of our language. It no longer provides us with a behavioral expectation. Certainly all dogs will still be four-legged. Certainly they will all bark, certainly they will all have nonretractile claws. But the application of the broader term "dog" no longer carries any significance whatsoever regarding domesticability, tameness, or adaptability to household tasks. In providing a language which will embrace all of our perceptions, we have deprived ourselves of a language which will tell us very much about what we see.

Between these two extremes there is an infinity of alternate choices which preserve precision. We need not draw our circles of precise meaning so small that they are restricted by our past experience, nor need we draw them so large that they exhaust the entire universe of possible perceptions. But the effect of these compromises is simply to combine in varying degrees the two extreme disadvantages we have already considered. To the extent that they are too large they deprive the application of a name of its behavioral significance. For some names, let me repeat, are not simply artificial and arbitrary labels. They are focal points about which crystallize our expectations about the perceptual worlds, and the act of naming a particular perceptual complex is a positive act. It is a statement of belief about the future history of the named complex.

It therefore appears that many names, as mediators of the perceptual world, are not precise and cannot be made precise without depriving them of their utility as mediators. Names are not that sort of thing. And from this polarization between precision and behavioral utility derives both the power and the danger of names and more generally of language.

But the statement that many words of natural language cannot simultaneously preserve their utility and possess a precise meaning has one very important exception which we have already observed. This exception is provided by the words of the ideal scientific text, which we discussed in the last lecture. There you will remember we discussed the ideal text as a purely formal structure in which scientific laws appeared as postulates determining certain of the formal characteristics of referenceless symbols. And this formal structure was combined with a meaning system which determined the reference in the perceptual world of these abstract symbols in terms of which the laws were stated. Experience or experiment entered in the choice of the particular postulate set and in the corresponding choice of a meaning-system. The remains of the structure was formal or conventional in the sense we discussed. And the effect of the formalization was to extract from the meaning system all those portions of the earlier commonsense concepts which were in the vague fringe of meaning, and to restrict the criteria determining the applicability of the term to a set about which our knowledge was sufficient to permit the construction of a postulate system.

In this process the meanings of words are changed, and narrowed. Frequently they lose their behavioral adequacy during the process. For the world of science, at least to date, is poorer than the world of everyday life. We cannot direct our behavior on the basis of scientific knowledge alone. But the loss of behavioral adequacy is compensated by a gain in precision and scope whose values are too apparent in the world about us to require further comment here.

But in practice we do not achieve this complete formalization of our text. And historically it appears extremely fortunate that we do not do so. We do leave vague meaning fringes on scientific terms, and our research is always conducted within the area determined by these vaguer fringes. It is in these areas alone that questions can arise as to established theories. The effect of full formalization is to make a theory impregnable except in so far as precise measurements may display deviations from the postulated laws. And this is not the most usual source of scientific advance.

If for example the phlogiston theory had been fully formalized, weight relations in chemical reactions would have been made totally irrelevant to it. Matter as it enters into the formal structure of the phlogiston theory does not retain weight as one of its necessarily associated qualities, any more than the precise and formal definition of dog can retain fur-bearing as a necessarily associated quality. Motion as formalized in Aristotelian physics is not an abstraction to which acceleration is relevant. Differences in acceleration do not formally correspond to different motions in the Aristotelian formalism any more than differences in the material of which the measuring rod is composed correspond to different distances in the formalization of Newtonian physics. And, as we have seen, it was because of the extraformal associations of matter and weight or motion with change of speed that these difficulties were recognized at all and became problems for these sciences.

Thus the vague and behaviorally determined meaning systems of natural language are one of the most important vehicles for what we have previously called scientific orientations. The area of stable meaning is an area of what we take to be certain knowledge. In this area no questions arise. The area outside our meaning system is an area which can be mediated neither by our language nor our perceptions. Prior to a shift of meaning systems no questions can arise here, either. It is only in the area provided by meaning fringes that scientific questions can arise and that scientific exploration can occur.

This exploration may proceed in one of two directions. It may result in increasing the scope and precision of the existing meaning system. And this is what occurred during the period which we previously described as the classic period in the development of a scientific orientation.[5] Or, in a more interesting sense, this exploration may result in the total destruction of the pre-existing meaning system. It may result in a rejection of the old criteria of meaning and the establishment of new ones for the same word and the same perceptual complex. Or it may lead to regroupings.[6] And this occurs in a period previously described as a crisis state in scientific development. The stage which terminates in scientific revolution. These are periods of disagreement and debate about meanings. They are periods in which the significance of equating different meanings with words whose application has previously been identical becomes apparent. They terminate with new precise criteria for scientific meanings and frequently with new central cores of meaning for natural languages. They are simultaneously destructive and creative of scientific orientation, behavioral worlds, and meaning systems.

So the study of language and of formalism brings us back to the recognition of the same pattern in the language process that we have discovered in our study of the history of science and from the psychology of perceptions. And I would conclude my remarks this evening by suggesting that these three facets of the search for knowledge can only be separated artificially as we have done here for the purposes of our discussion. Whether we are scientists or laymen these three are inextricably intermingled in our daily practice. They represent simplifications of the flux of perception based on experiences; they embody our implicit knowledge; and within them we find our science. Our linguistic apparatus, our involuntary yet alterable organization of our perceptions provide us with our science in embryo. They are the vehicles of that inevitable predisposition to theories of a certain sort which, as we have noted again and again, govern our experiments and the conclusions which we draw from our experiments. By increasing abstraction and increasing precision we can create within the pre-existing patterns of language and perceptions a summary of our most certain knowledge which we call science. And in the process we can gain some additional knowledge. It is this which we embody in scientific texts. But it is not knowledge of a different sort, nor is it gained by an essentially different procedure, from the more primitive experience embodied in the organization of our language and perception. Without this more primitive organization, we cannot proceed to act or to do research.[7]

But, and this is the source of the difficulty which we have encountered again and again during the course of these lectures, our perceptions and our language which provide this underlying organization are not permanent and immutable. They arise from experience, they legislate for experience, but they may prove inadequate to experience. When they do, they must be altered, and the process by which they are altered is destructive as well as constructive. Thus the patterns of organization from which science proceeds limit it. Without them there would be no science. Within them there can only be certain sorts of science. So continuing progress in research can be achieved only with successive linguistic and perceptual re-adaptations which radically and destructively alter the behavioral worlds of professional scientists.

Notes

1. Here Kuhn instructs himself, "go to board." The illustrations he offered his audience are evidently similar to those he later published in his essay "Second Thoughts on Paradigms," in *The Essential Tension* (Chicago: University of Chicago Press, 1977), 293–319, 311.

2. See, for example, Stuart Chase, *The Tyranny of Words* (New York: Brace, 1938).

3. Months after Kuhn wrote this, a similar semantic criticism of anticommunism appeared in the philosophical literature. See Victor Lowe, "A Resurgence of Vicious Intellectualism," *Journal of Philosophy* 48(14) (1951): 435–47.

4. Here Kuhn directed himself to the blackboard.

5. In Lecture V, Kuhn called it the "classical period."

6. Kuhn inserted this sentence by hand, and place it in brackets.

7. Kuhn inserted by hand a short, illegible word at the start of this sentence, possibly "But" or "Thus."

APPENDIX

"The Prevalence of Atoms": Kuhn's Original Outline*

Notes: Lowell Lecture III: 3/9/51
THE PREVALENCE OF ATOMS

I. INTRODUCTION
1. Last lecture studied the transition in the attitude of Western scientists and philosophers toward a single problem: that presented by the study of motion of terrestrial bodies.
    a) More precisely—examined two different points of view toward a single problem, motion, and showed how particular limited inadequacies in the first of these had led to a complete reformulation of the problem and to a new set of laws.
2. Tonight I should like to attack our problem somewhat differently. Instead of a single problem, we shall examine the application of a single approach to a variety of problems:
    a) Manner in which one metaphysical notion about the structure of the world has provided new insights to science and scientists.
3. This notion is the one commonly called ATOMISM.
    a) Belief that world is made of infinite number of microscopic particles in constant motion in an infinite void.

*This transcription of Kuhn's outline for his third lecture, "The Prevalence of Atoms," reproduces his typed outline with basic corrections of grammar and spelling, expansion or regularization of Kuhn's abbreviations, and typographical corrections shown in brackets. His outline includes three marginal notations not included here: "Omit if past 2[?] min." at §V, 6–7; "Omit if necessary" at §XI; and "read" at §XII, 6.

b) And this idea about the structure of the world is one which, AT LEAST HISTORICALLY, we owe to the Greek philosophers Leucippus and his student Democritus who flourished in the 5th century B.C.
4. I say, "at least historically," for I should like this evening to find a middle ground between two extreme views which have been held regarding the relation of modern scientific atomism to the atomism of the Greeks.
   a) First is the truly absurd view that the Greeks, by sheer power of mentation, anticipated many or most of the conclusions produced by the combined efforts of XIX and XX scientists.
   b) Second view is equally absurd. States that resemblance between Greek and modern atomism is purely fortuitous. That it's simply a rather disagreeable accident that a lucky guess by some obscure Greek philosophers should have seemed so similar to the totally different theory provided by careful experimental and mathematical research during the past 150 years.
5. As an alternative—the question is misphrased—science since beginning of XVII cent. has believed in and made use of a number of different atomisms.
   a) There is no one scientific atomism.
   b) Different fields of science have occasionally employed incompatible atomisms at same time.
   c) But these various atomisms have evolved through the attempt to employ an older atomism in a new scientific context. And in this attempt the basic notion of an atom has itself changed.
   d) But if atomism has not been unaltered in its contacts with science—neither has science. The continued association of science and atomism has been fruitful. If we trace the matter back.
   e) And the first philosophical atomism to which science owes an important debt is that of the Greeks.

II. GREEK ATOMISM

1. Greek atomism in large part a produce of the same set of dialectic problems examined at beginning of last lecture.
   a) Parmenides: everything which exists is eternal and changeless.

b) Thus the vacuum, whose very name implies the absence of all being, cannot exist. For how can non-being exist.
   c) So there is only one thing, the real universe. And it is completely full, self-contained, and forever changeless.
   d) This is first of the historically important phil. monisms.
 2. Aristotle's response. Agree that there's no void. Universe is full.
   a) But fullness doesn't imply absence of change—motion can occur by anti-peristasis. Motion of the fish.
 3. This answer dominated thought in Europe until well into XVI century. Not surprising. Built into physical system giving commonsense world.
 4. But it was not the only way of preserving change from the onslaught of Parmenides.
   a) Alternative provided even before Aristotle.
   b) And this was contained in Leucippus's denial that there was anything self-contradictory in the notion of the void.
   c) On the contrary: the void exists, and so do little bits of microscopic matter which move about in it.
   d) The flux of appearance is provided by the motions of these corpuscles.
 5. but for dialectic reasons similar to those which led Parmenides to keep the all indivisible, the Greeks held that these particles could not be divided.
   a) Atom means undivided.
   b) Thus of the Parmenidean monism they made a monadology.

III. PHILOSOPHICAL CONSEQUENCES
 1. This view of the nature of reality had number of psychological consequences of great importance for later scientific thought.
 2. There are holes in nature. A vacuum can exist.
 3. The universe is infinite. Extends forever in all directions.
   a) Contrast with ARISTOTELIAN UNIVERSE bounded by finite sphere.
   b) For what can bound it except atoms & the void.
 4. More important—all the changes, all the variable qualities we observe in nature are produced by the changes in positions and relative motions of the fundamental particles.

a) In the words of the later atomist Epicurus: "We must suppose that the atoms do not possess any of the qualities belonging to perceptible things, except shape, weight, and size, ... For every quality changes; but the atoms do not change at all, since there must needs be something which remains solid and dissoluble at the dissolution of compounds, (something) which can cause change; . . . changes affected by the shifting in position of some particle, and by the addition or departure of some other."

5. This is a very interesting statement, for the ideas which it expresses have been a continuing source of two fundamental ideas about the universe, which have been a continuing source of both inspiration and trouble for later sciences.

   a) The atoms are sensuously neutral. They possess only size & shape: i.e. extension.
   b) Thus to understand the variety and sensory luxuriance of the world exhibited to our senses we must study not these sense impressions themselves, for these are scarcely trustworthy, but we must search for the arrangement & motions of the corpuscles which underlie the appearances.
   c) This is beginning of the tremendously important division between primary & secondary qualities.

6. Even more important is the notion of the world as a machine which lies behind the description of the atomistic universe.

   a) All the universe is made up of the same sort of stuff.
   b) Once the atoms are given their initial motions they go on by themselves. After its start the atoms go on by themselves, under their own laws.
   c) This is notion of universe as a giant machine, a piece of clockwork.

IV. DEPENDENCE ON THE GREEKS

1. This idea of the universe as giant machine is frequently thought to be a consequence of the Newtonian philosophy which so dominated XVIII century thought.

a) Newton thus described as a man who single handed gave the world notion of world as a machine—a notion in which he did not entirely believe himself.

2. This is not far wrong if we are concerned with wide popularity of the notion. With the notion as thoroughly dogmatic belief which influences not just science and philosophy, but also politics and art.
3. But for many working scientists notion is 3/4 of a century older than publication of PRINCIPIA.
    a) And here its source is unequivocally Greek.
4. At beginning of the century FRANCIS BACON had written a defense of the works of Democritus and other Greek atomists.
    a) Throughout his life he proclaimed that one of the fundamental tasks of the new science was study of motions which underlie the qualities present to our senses.
    b) Illustrated both in his own abortive researches, chiefly heat, and with many examples drawn from Roman atomist Lucretius.
5. Only slightly later Rene Descartes, French philosopher whose work so influenced the continental science of the XVII century, provided a complete model of the world as a machine, governed by invariable God-given laws.
    a) And the building blocks of the Cartesian universe were little corpuscles, acting only by impact.
    b) In fact, Descartes in the beginning was accused of simply cribbing his philosophy from the Greek atomists.
6. I will not multiply names, but the notion was extremely prevalent.
    a) In fact the very name which was attached to the major scientific tradition of the XVII century—"the New Philosophy" or "the Mechanical Philosophy" seems in large part to have denoted an attempt to reduce all phenomena to the motion of such elementary corpuscles with the aid of experimentation.
7. ROBERT BOYLE, in so many ways the leader of the movement for the New Philosophy, etc. was admittedly indebted to the Greek atomists.
    a) Describes universe as "a self-moving engine," "a great piece of clockwork."

b) His major work is an attempt to apply these notions to chemistry.

V. SOME EFFECTS ON PHYSICS

1. So far have dealt with atomism as purely speculative cosmology.
    a) Drawn by free creative power of human mind from the consideration of logical or pseudo-logical problems.
2. To such speculative, cosmological thinking you may wish to deny the name of science. May wonder why I go into it in lectures devoted to the study of problems in scientific method.
3. I think you'd only be partly correct in denying the name of science to this material. But I am not now prepared to argue it. Reserve this discussion for fifth lecture.
4. Now just want to point out that whether it is science, it has important effect on researches that are indubitably scientific.
    a) Now turn to an examination of some examples.
5. First of these is drawn from material of last lecture: Foundations of Dynamics. Should like to point out an important effect of the recovery of a limited portion of the atomistic viewpoint upon the study of motion.
    a) Aristotle: Universe is full; motion is through plenum.
    b) Thus always two forces involved in a motion. Pusher & medium. This holds until the XVI century.
    c) But XVI cent., though not marked by scientific adherence to full-scale atomism of Democritus, does show a large-scale rejection of Aristotelian plenum.
    d) Many scientists believe on philosophical grounds that there are tiny vacua in things. These account for condensation & rarefaction. Very light substances like air are highly rarefied. They are almost all vacuum.
    e) Thus motion thru air is very like motion thru a vacuum and this suggests that motions in which there is no resistance are profitable subject for scientists to consider.
    f) Motion under one force becomes a worthwhile abstraction.
    g) This is the background for the new acceptance of Galileo's argument about the two bricks which had been tried and rejected at times when the medium was thought to play a different role.

      6. But effect of new attitude toward the void is not felt just in dynamics.
- a) Idea that there are vacua and that their size can be altered by finite forces suggests a new problem presented by an old phenomenon: the problem of pumps.
- b) Known for more than a century before Galileo's time that pumps in mines wouldn't raise water large distances. Generally less than 30 feet.
- c) Not surprising—pump shaft of wood etc. crude boring.
- d) People continued happily to design, on paper, pumps which raise water hundreds of feet.
- e) Idea that there's an inherent limitation here doesn't arise. Can be no vacuum, so if plunger is drawn up water must follow in absence of leaks. And the leaks certainly existed.
- f) Faced with this situation today we'd eliminate most of the leaks, discover that this scarcely increased pump effectiveness.
- g) Galileo can't do this very well. Thirty-five foot leak-proof tubes can't be fabricated.
- h) But even without them, believing that a vacuum could conceivably be made by a finite force he is able to suggest that perhaps that's what's happening in the pump. Water breaking away from plunger under its own weight and leaving vacuum.

      7. Parallel to pendulum: new point of view changes significance of an old observation. It aides in the creation of a new problem.
- a) Doesn't do this single-handed. Several generations of trying to improve pumps also an important factor.
- b) And it doesn't solve problem. Galileo was wrong. Correct explanation awaits Torricelli and mercury barometer experiment. But Torricelli was a pupil of Galileo's, and his experiment is a consequence of Galileo's isolation of the problem.

VI. EFFECTS OF THE FULLER ATOMISM

1. Need not restrict ourselves to this partial atomism, which, since it does not hold for fundamental particles perhaps ought not be called an atomism at all.
2. The effects of the full atomism of Democritus are equally clear.
3. For example, Galileo had suggested that a ball rolling on a horizontal plane will continue with a constant velocity in a straight line forever.

- a) This is very close to what we now call the principle of inertia, but it misses it. For it holds only for horizontal plane.
- b) Not surprising. Can't get this notion from experiment; balls roll slower up and faster down inclines. Only on horizontal planes do they go on for some time. Even here can't get rid of friction. So the law is an idealization and an unlikely one for Galileo because—
- c) Galileo is still in many respects Aristotelian. He has rejected natural motion vs. violent motion distinction. But horizontal vs. vertical motion still seems different. UP & DOWN VS. SIDEWAYS. Motions are relative to earth.
- d) But in atomistic universe there are no UPS & DOWNS. Fundamental motions are those of elementary corpuscles, swimming in an infinite void. And an infinite void has no directions.
- e) Atomist Descartes first enunciates principle in full generality, with background of sorts of considerations which lead Galileo to limited case.

4. But atomism leads Descartes still further. It provides him with an entirely new sort of dynamical problem.
   - a) Since all change of this linear motion is through impact, we must study impact. What happens when two elementary particles collide? How is motion transmitted from one to the other.
   - b) Descartes enunciates bad laws.
   - c) These are corrected later. Lead to conservation of momentum.
   - d) Problem is not in literature before. It has no apparent importance. Dynamics was something else.
   - e) Not it is almost the pre-eminent dynamical problem. The problem of billiard balls. But it didn't arise that way.

5. Examples of this sort could be multiplied almost indefinitely.
   - a) Not only dynamics, but Heat, Light, Chemistry.
   - b) Instead let me simply remark on the extent to which these developments culminate in and are given a new form by the work of ISAAC NEWTON.

6. Have already suggested one sense in which atomism provided a motive and a direction to Newtonian research: World Machine.

a) Universe like watch, composed of same sort of material everywhere, and these materials operate in essentially same manner throughout.
   b) This as much as watching the apple, lies behind search for "universal gravitation" for a force which would be the same for the apple and moon.
7. But adjective "universal" has still another meaning which is, I think, even more thoroughly atomistic.
   a) We learn universal gravitation as attraction inversely proportional to distance between center of bodies.
   b) Newton pronounced himself dissatisfied with this formulation. He demanded and found a proof that if all little particles within a body acted this way then total effect would be the same as though entire mass of larger body were concentrated at its center.
   c) But this demand that forces acting between heavy bodies be the net results of identical forces acting between the microscopic bits of matter which compose those bodies is a new one. Kepler etc. quite content to find forces acting between bodies themselves.
   d) This insistence upon force laws which work between bodies is a new one; and I think its source is in the notion of an atomistic world machine. Descartes, whose system was not successful, had imposed same demand, but I know of nothing which parallels it before the popularity of atomism.
8. Newton's fruitful application of atomism not restricted to the Dynamics.
   a) Boyle's law
   b) Optics: reflection, refraction, and simple diffraction, etc.
9. So after Newton everyone was an atomist.
   a) But a new sort of atomist. Elementary particles moving under influence of forces acting between them.
   b) And the notion of forces acting at a distance was new and radical. Until this time only impact had produced change.

10. This was new: Led to notion that object of science was to discover new sorts of forces between bodies, and to determine the effects of such forces.
    a) Much BUT NOT ALL of the science of the XVIII & XIX centuries is given motive and direction by this new sort of atomism which itself derives from the Newtonian synthesis.

VII. TRANSITION TO DALTON

1. Since Newton's day there are too many fruitful applications of atomism & too many modifications of atomism to permit our continuing even so superficial a sketch. Too much & too technical.
    a) Instead will reserve for the end of the hour a few more general remarks on the subject of modern atomism.
2. Will now try a more detailed illustration of way in which atomistic ideas to a particular set of problems can suggest new significances of old data, and of manner in which atomism is itself modified by the application.
3. For such a purpose could scarcely find more central or typical figure than English chemist John Dalton.
    a) His work at beginning of last century brings genuine atomism to chemistry.
4. Better say shows fruitfulness of atomism for chemistry, for even before his time a number of influential chemists believed that substances with which they dealt were built up of atoms.
    a) This was an atomism borrowed directly from the physicists of the period and its use was to explain that group of properties of natural substances which we should now call physical.
5. DRAW PILE OF SHOT MODEL ON BOARD
    a) Pile of shot model: space filling atoms
    b) The individual particles are themselves composed of two different substances. Core & Caloric.
6. DRAW CORE & CALORIC sheath.
    a) Core accounts for weight & nature of substance.
    b) Caloric is weightless fluid providing a jelly in which particles are set. It provides the forces between them.
7. Utility of model.
    a) Difficulty in compression.

b) But expansion by heat—and Caloric is just heat.
   c) So uniformity of expansion on heating.
   d) And change of state.
   e) And different heat capacities.
  8. Not quantitatively useful—but reduced many divergent phenomena to order.
      a) But not chemical phenomena. No notion that these particles enter into reactions a such.
      b) Hydrogen particles, Oxygen particles, Water particles. But this is static model.

VIII. DALTON'S PROBLEM & ITS SOLUTION

  1. John Dalton, man who bridged this gap, was not a chemist but a physicist. Deeply influenced by Isaac Newton's atomism.
      a) More precisely his own work was almost entirely in meteorology and heat.
  2. As a meteorologist he was much concerned with a chemical discovery made while he was in his teens.
      a) Air is not simple but compound. It's a mixture of two gases—Oxygen, Nitrogen. Plus some water vapor.
  3. This raises very serious difficulties.
      a) Two gases are different in weight—why don't they form strata? Heavier Oxygen at the bottom.
      b) And problems of Gas absorption.
  4. Dalton's attempts. Different ways of stacking—different sizes for Caloric sheaths, etc.—different sorts of force laws.
      a) Did convince himself that no one force law could account for lack of stratification and absorption etc.
      b) But managed to convince himself that force law could be worked out if and only if atoms (with sheaths) were of different sizes and weights.
      c) No point in examining details of theory. It would not have worked—in fact it's unusually absurd—but it provided a new problem. Find sizes and weights of the ultimate corpuscles.
  5. Dalton's great genius lay in pointing out that this could be done on the basis of existing chemical data.
  6. The assumptions:

a) Substances made of atoms.
   b) All atoms identical—otherwise we'd have two sorts of Oxygen, etc.
   c) Then with reasonable assurance can compute relative weights and volumes of the fundamental particles.
7. Example
   a) 8 pts by weight of Oxygen combine with 1 pt by wt. Hydrogen to form water.
   b) If this is one particle to one particle, then weight of Oxygen is just eight times that of Hydrogen, because . . .
   c) Could be two atoms to one—then weight relation would be sixteen to one. But wisely refuses to complicate the situation more than is required.
8. By examining other compounds gets weights of other atoms, and not only weights but also relative volumes of the corpuscles.
   a) I[f] atoms are space filling and you know relative densities and relative weights of fundamental particles, can discover relative sizes.
9. So he got a whole list of relative weights and relative sizes of the atoms. The sizes were different and he convinced himself that his mechanism for the atmosphere would work out. HE WAS DELIGHTED.

IX. DALTON AND THE CHEMISTS

1. If this was all that there was to Dalton's theory no self-respecting chemist would have paid any attention.
   a) Atmosphere mixing wasn't a very big problem, and it took very little perspicacity to see that Dalton's theory of the forces between the particles wouldn't really account for anything.
   b) The notion of atoms uniting with each other to form new atoms was pretty thoroughly speculative. And in any case why should they unite this way.
   c) Why not 2 Oxygens to one Hydrogen or conversely, or 7 to 11, or something else.
2. But though Dalton had designed his theory and done his work to take care of atmosphere and absorption etc. it turned out that his

speculation, useless for his own problems, could do a great deal for chemists.
3. For example, Great debate about whether substances can combine in any old proportions.
    a) Hydrogen and water show only one proportion. But copper and oxygen seem to combine in almost any proportion you choose.
    b) If Dalton was right there were only certain proportions in which elements could combine. This meant that others must be mixtures.
    c) Thus had a clear criterion for chemical vs. physical & this was of very great importance.
4. But Dalton's work had other consequences even more striking for chemists, and I should like to illustrate one of these.
5. Dalton had applied his method to substances which combine in more than one ratio to form more than one compound.
    a) Many such reactions had been known before his time and the analytical data was available to him.
6. Dalton suggests that for such cases one compound must be binary and two ternary—
    a) Application to the nitrogen oxides—ON BOARD
7. Similarly for the oxides of carbon, etc.
8. Thus law of multiple proportions: A given amount of one element will combine with weights of a second element which bear to each other simple whole number ratios.
9. But the data for this had been available for years.
    a) Just at Dalton's time a number of chemists were about to announce the existence of such a regularity for a few particular compounds.
    b) Dalton supersedes them—by showing mechanism and generality. Many of them don't even publish their results.
10. So a new law has entered chemistry and a new guiding principle to be used in all chemical manipulations, for a completely irrelevant source.
    a) For gas problems with which Dalton was concerned were not relevant to chemists and Dalton hadn't solved them.

b) The central parts of the discovery from the point of view of chemists were asides for Dalton. They had relatively little significance to the central portion of his work, which was, at least initially, the search for forces between the fundamental particles.

X. DALTON & GAY-LUSSAC

1. Just how irrelevant Dalton's considerations really were is shown by an immediate sequel to his discovery. A sequel which actually destroyed the theoretical underpinning of his discovery.
2. In 1809, just two years after Dalton's announcement of his theory, the French chemist Gay-Lussac discovered another regularity of chemical reactions.
    a) Law of combining volumes. Illustrate.
3. This was immediately accepted with rejoicing by all chemists. And they believed it a beautiful proof of the atomic nature of their substances and their reactions.
4. But Dalton couldn't believe it. He not only doubted its generality, but he accused the brilliant experimentalist Gay-Lussac of twisting his figures to produce the apparent regularity.
5. This isn't strange.
    a) Dalton's insistence upon differences in volumes meant an insistence upon different numbers of particles per unit volume.
    b) But simplest interpretation of Gay-Lussac was equal numbers of particles or at least integral ratios of numbers of particles. And this wouldn't fit Dalton's data at all.
    c) Within two years of his proclamation of it the theory was destroyed. Though he held out until the end of his life.
6. But it wasn't destroyed for Chemists.
    a) They weren't committed to his theory of forces or to his computations of sizes.
    b) They simply rejected the part for which Dalton had done the work and held on to his asides.
    c) And they drew and still draw great profit from this.

XI. AVOGADRO'S HYPOTHESIS

1. But the chemists were still not out of difficulty.
2. Examine Hydrogen and Oxygen reaction. Must mean H_2O.

3. But then where do the extra Oxygens come from for the extra volume of steam?
4. Two ways of solving this problem.
    a) Only half as many water particles per unit volume. This works, but it leaves a certain arbitrariness in chem formulae. Perhaps then the same is true of Hydrogen and water is still HO.
5. A better method was suggested by Italian physicist Avogadro two years after Gay-Lussac's discovery.
    a) Preserve equal particles for equal volumes.
    b) But these particles aren't ultimate. They can be broken down in chemical reactions. APPLY TO WATER.
6. This works fine, but it's absurd.
    a) When is an atom not an atom?—"atom" means undivided.
    b) If you can break them in two, why not in three, for five, or into a hundred fragments? What is the good of the notion at all?
7. In fact Avogadro's proposal was scarcely even taken seriously for almost fifty years.
8. Can't retrace the story here—but would like to note for future reference that
    a) Rejection caused great difficulties—Everyone wrote atomic weights and molecular formulas differently. What was the formula of water and what the weight of Oxygen.
    b) The result was very serious difficulty for all of chemistry. No agreement in formulas on nomenclature.
    c) By mid-century as the number of known compounds proliferated the situation became almost unbearable. It became difficult for one Chemist to read another's paper without a table of the atomic weights used by that chemist.
9. As a result atomic theory itself was almost abandoned by many chemists.
    a) And with the theory itself shaken to its foundations, it was a great deal easier for chemists to relinquish the inviolability of the atom in order to gain a unanimity in their tables of weights and formulas.
    b) This return to the modern view was finally accomplished in 1858, by Cannizzaro.

10. But this again meant a new sort of atom.
    a) It could be broken.
    b) it was not space filling—no pile of shot.
    c) No caloric sheaths to explain the physical properties so nicely.

X[II]. CONCLUSION

1. Clearly these chemical atoms of the late years of the XIX century were different from any we've met so far.
    a) They were not all made of the same sort of stuff as the atoms of Democritus were.
    b) They could be split up in chemical reactions.
    c) Nor did they have any function in explaining the physical properties of the substances into which they entered.
2. The function of explaining physical properties was taken over at this time by a quite different sort of atom developed by the physicist.
    a) The atom of the kinetic theory. Very small compared with space for it in a gas.
    b) In continuous rapid motion through vast empty space, etc.
    c) Hard little ball which rebounds with perfect elasticity when it hits another atom.
3. In fact the possibility of reconciling the two sorts of atom seemed so remote that there was at the end of the century a widespread rebellion against all atomism.
    a) It was admitted that atoms might be useful devices to think in terms of.
    b) But it was said that science had shown the notion of such fundamental particles from which the universe was constructed to be absurd.
    c) Believing in them, we were as bad as the Greeks for we were indulging in the same sort of futile speculations.
    d) Useful or not in the past, the science of the future should purge itself of such notions.
4. Fortunately the movement was not successful
    a) In the past twenty-five years and only then we've succeeded in reconstructing an atomism which will account for both physical and chemical properties of the atoms.

www.ingramcontent.com/pod-product-compliance
Lightning Source LLC
Chambersburg PA
CBHW031926240526
45464CB00023B/1689

  b) It will even tell us a good deal about the qualities of the atoms in an aggregate.
5. But once again it's a new sort of atom.
  a) The planetary atom—electrons and a hard nucleus.
  b) And nucleus itself can be split into an increasing number of different sorts of particles.
6. What our notions on the subject will be fifty years from now is almost impossible to predict—but they will be different.
  a) And the new ideas about fundamental particles will be achieved by the only tools we know how to use:
  b) The application to the new problems of the nucleus of our existing and inadequate conceptual tools gained from an examination of the planetary atom.
  c) For just as in preceding reshaping of atoms, it is in the course of such application of our old tools to new problems that the discovery which will reshape our notions of atoms and of the nucleus will be made.

FURTHER READING:

If you enjoyed this book, please consider reading one of the other books in the series:

Making Money Online: Book 1 (Understanding the Online Landscape)

Making Money Online: Book 2 (E-commerce and Online Retail)

Making Money Online: Book 3 (Freelancing and Remote Work)

Making Money Online: Book 4 (Content Creation and Monetization)

Making Money Online: Book 5 (Online Tutoring and Education)

Making Money Online: Book 6 (Online Surveys, Microtasks, and Rewards)

Making Money Online: Book 7 (Online Investments and Trading)

Making Money Online: Book 8 (Creating and Selling Digital Assets)

Making Money Online: Book 9 (Online Consulting and Coaching)

Making Money Online: Book 10 (Maximizing Online Income Opportunities)

All the books can be found on Amazon as Kindle and Paperback, or you can buy the complete edition which contains the full series in one book. The complete edition is available as Kindle, Paperback and exclusively as Hardback. You can find all the links in my book site: books.michaelmayaka.co.uk.